Monographs on Astronomical Subjects: 3

General Editor, A. J. Meadows, D.Phil.,

Professor of Astronomy, University of Leicester

Cosmology and Geophysics

In the same series published by Oxford University Press, New York

Cosmology and Geophysics

Paul S. Wesson

St John's College
Cambridge

Monographs on Astronomical Subjects: 3

OXFORD UNIVERSITY PRESS
NEW YORK

© P. S. Wesson 1978

First published in the United Kingdom 1978 by Adam Hilger Ltd, Bristol
Adam Hilger is now owned by The Institute of Physics
First published in the United States 1978 by Oxford University Press, New York
ISBN 0 19 520123 X

Printed in Great Britain by Adlard & Son Ltd
Bartholomew Press, Dorking, Surrey

Preface

This book is concerned with those cosmologies that have salient consequences for geophysics, and those theories of gravity which can draw observational support or censure from certain non-Newtonian processes observable in the Earth. The majority of these theories are concerned with a possible time variability of G, the Newtonian gravitational parameter. Attention is also given to certain (not widely appreciated) effects in astrophysics and geophysics of Einstein's general relativity. The account given should be of interest to those concerned with the implications of modern cosmology for geophysics and astrophysics, and to those concerned with geophysical subjects, such as the rotation and long-term geodynamical evolution of the Earth, that can be seriously affected by cosmological phenomena. While most of this account is intended to be a survey of the field, some of the argument has been influenced by my own researches (not all directly connected with the meaning of G) carried out mostly in Cambridge over a period of several years. During this time I have had reason to be grateful to many people for checking results, offering comments and personally communicating information with which I would not otherwise have come into contact. I would therefore like to acknowledge W B Bonner, C Dyer, C Gilbert, R N Henriksen, members of the Institute of Astronomy, Sir H Jeffreys, S J Knight, A Lermann, R A Lyttleton, A J Meadows, A A Meyerhoff, J V Narlikar, M Rees, M G Rochester, M Rowan-Robinson, I Roxburgh, S K Runcorn, W C Saslaw, R Stabell, J L Synge and R Tavakol.

The bibliography is as complete as I can make it, and the survey is meant to be comprehensive. To avoid confusion, I may mention here that throughout the text I have used units most familiar to the fields being discussed. The symbol ($\simeq$)

means equal to within a factor of three, $(\approx)$ means equal to within an order of magnitude, while $(\sim)$ means expected to be asymptotically equal. In cosmology $(\approx)$ sometimes has to be taken in a more liberal sense.

Paul S. Wesson
St John's College
Cambridge

Contents

1. *Introduction*

Ancient cosmologies, as systematic if unscientific systems of philosophy, were to a large extent synonymous with the study of the Earth, especially with the relation of the latter to the Sun and other planets. Cosmology as a science, beginning at the start of the twentieth century, rapidly widened its scope of inquiry out to the size of the light sphere, which is of dimension 10^{28} cm and so is considerably larger than the characteristic scale 10^9 cm of geophysical studies as they concern the Earth. The extent of the Universe may be vast as judged by the size of the Hubble sphere, if indications from the mass field theory of Hoyle and Narlikar are substantiated. It is a remarkable property of the cosmology noted, along with those of Brans and Dicke, Dirac and others, that formalisms deduced on the astrophysical scale have unavoidable consequences for processes on the geophysical scale that can be estimated and compared with observation.

The three above-mentioned cosmologies can all be developed in a way that makes the Newtonian gravitational parameter, G, time-dependent, and possibly also space-dependent. These and other theories therefore directly affect geophysics via the effects they have on the rotation of the Earth, the orbital motion of the Moon, the expansion of the globe and the influence of these phenomena on geodynamical processes, such as sea-floor spreading. While rotation and expansion have long been realised to be possibly connected with cosmological considerations, the other main field of geophysics that has usually been regarded as possibly cosmically linked, namely the origin of the Earth's magnetic field, is not now believed to be so strongly implicated, except indirectly (Wesson 1973). It is observed to be the case that increases in size, decreasing rates of rotation and the decay of magnetic fields are phenomena which are seen

1

together in many stars (Wilson 1966), and the picture of a gradual running down, into which the Earth may also fit, is a familiar one. Venus and the Moon rotate slowly and have negligible magnetic fields, but with the abandonment of the Schuster–Wilson hypothesis (namely, that all massive bodies have magnetic moments proportional to their angular momenta; see Blackett 1947), there is no known way to connect these phenomena in a direct manner. This situation is not unsatisfactory, since the theory of the geomagnetic self-exciting dynamo can adequately describe the origin of the field (if with a large latitude of uncertainty as to its precise mode of functioning).

The same cannot be said of most other aspects of geophysics, despite the fact that a large body of data exists that can be interpreted on the basis of several well known geodynamical hypotheses. In particular, while there is a general feeling that convection occurs in the mantle, there is no consensus on its location, and certain workers (Jeffreys 1970, Wesson 1972a, Meyerhoff and Teichert 1971, Meyerhoff and Meyerhoff 1972) have criticised a too-ready acceptance of simple plate tectonic models on the ground that the physical processes on which they are based are still relatively unknown. This is true in the present context, where it is clear that certain well founded cosmological theories, such as those of Dirac (1938, 1974), Brans and Dicke (1961) and Hoyle and Narlikar (1971), predict substantial amounts of planetary expansion that are often not taken into account in constructing geodynamical models. They also predict, by implication, effects on the geochemical and geomagnetic evolution of the Earth's interior and modifications over geological time scales of the angular momentum balance in the Earth–Moon system.

It would seem wise at least to consider such effects in the construction of geodynamical models of the planet. Wesson (1972a) has listed 74 points which require further attention in the field of contemporary global tectonics; one main point is that geodynamical models should not rely on whole-mantle convection as a driving agent, while another is that secular global expansion may be a continuing major process in the Earth. There is, of course, no objection *a priori* to expansion and whole-mantle convection, since as Holmes (1965) put it,

2

'Convection does not exclude global expansion. Global expansion does not exclude convection'. It should, however, be borne in mind that the correctness of a wide range of non-Einsteinian cosmological models would imply the acceptance of expansion, and not convection, as the primary geodynamic process.

In the reciprocal sense, if certain geophysical processes can be confirmed as operative in the Earth, then the implications for cosmology are profound. The variability of G and the other parameters of physics, the metric dilation of space and the origin of inertia are all subjects that stand to be elucidated by studies of the Earth and the lunar orbit. While the reference background for nearly all modern cosmologies continues to be Einstein's general relativity, with the concomitant importance of the Robertson–Walker line element and the Friedmann models of the Universe, departures from this basis would be drastic if refined data were to confirm the existence of some non-Newtonian geophysical effects connected with geodynamics and the secular deceleration of the planet's spin.

Since general relativity is the standard theory which is used to judge most of the new cosmologies discussed in following chapters, this is an appropriate place to make some brief comments for those who are not familiar with Einstein's theory. The latter is a tensor theory of space–time, which means that it employs a notational convention for grouping together a set of nonlinear partial differential equations in the space and time coordinates. The tensor notation should not be viewed as complicated; in most practical cases a tensor equation readily splits up into a set of only a few (not too complex) differential equations. The canonical set of these is presented and employed in Chapter 7. A good review of tensor calculus and general relativity is given by Lawden (1967). Tensor notation is used in following chapters for the sake of brevity, but only in places where it is required for conciseness. The metric tensor is a notation for the set of parameters that expresses the distance between two points in four-dimensional space-time, this distance being the interval. The motions of particles are described by geodesics which are most often found by varying an action (that is, an energy) to find the equilibrium trajectories of particles in the space–time. The choice of an

3

action and equations for the theory depend on certain principles, the most important of which for general relativity is the Principle of Equivalence. In its strongest form, this says that at a given point all laws of nature have the same form as in special relativity when expressed in local coordinates. This principle has been much discussed and is often used in a variety of weaker forms. In practice, it means that bodies moving in force fields due either to an acceleration or to the presence of a massive body behave in the same way. Another principle that Einstein tried to incorporate into general relativity was that due to Mach. This implies that inertia (the resistance to acceleration) of local matter is due to the rest of the matter in the Universe, but Mach's Principle is hard to define exactly and is generally believed not to be incorporated fully into general relativity. It is probably a cosmological property of matter. On the cosmological scale, the solutions to Einstein's equations of motion that are spherically symmetric about a point are the easiest to study. Starting off with an impulse from the centre, a sphere of matter either continues to expand forever outward (a hyperbolic or open model); expands infinitely, but gradually comes to rest (a parabolic model); or expands to a maximum and then re-collapses (an elliptic or closed model). Space–time is termed flat if the model is parabolic and Euclidean at infinity, and curved otherwise. Often in these types of behaviour, a randomly placed observer cannot see all the Universe because natural barriers or horizons are present over which causal influences cannot operate to reach the observer. The advancing front of a sphere of light is the most simple type of horizon. The expanding or contracting motion of a cloud is usually described by a number with the dimensions of an inverse time. For the Universe, this is Hubble's parameter, H; its reciprocal is a rough measure of the age of the model. Another parameter that may be of great importance in cosmology is the cosmological constant, Λ. This represents a force different from normal gravity, which is trying to push matter apart (if it is positive), or pull matter together (if it is negative). The cosmological constant is a controversial parameter, but it will be seen that it is really a term that describes those parts of the model where there is no matter (namely, the vacuum or empty parts of the space–time). The special class of spherically

4

symmetric solutions to Einstein's equations that have uniform density (are isotropic and homogeneous) are the Friedmann models, which have Robertson–Walker metrics. All other models which might be of practical importance are often termed hierarchical, and are anisotropic and inhomogeneous. While the effects of theories of the Einstein type are dominant on the cosmological scale, it should be remembered that their effects can also be seen as departures from Newtonian physics on the scale of the Solar System and even smaller bodies, and can in principle be important on all scales from that of the horizon of the Universe down to that of the atom.

In order to explain the interrelationship of cosmology and geophysics, the following plan has been adopted. Chapter 2 gives an account of those cosmologies in which G is variable, and examines the tests and limits for such variability set by geophysics and astrophysics. A discussion of the more general aspects of those cosmologies as they concern gravitational theory is deferred to Chapter 3, which is mostly concerned with the formal bases of (mainly metric) theories of gravity. Chapter 4 examines the status of the very large, dimensionless Eddington numbers that appear in many theories and in numerous empirical situations in nature, paying particular attention to G and its possible variability. Chapter 5 reviews the subject of the Earth's rotation and how it can be employed to obtain limits on the time variation of G and other cosmological effects such as expansion. Chapter 6 is an evaluation of the expanding Earth hypothesis. This is followed in Chapter 7 by an account of expansion in relativity, showing how one can assume equations such as those of Einstein to hold in some gauge of a metric theory of relativistic gravity, impose conditions on the equations and so explain certain cosmological effects as they are manifested in geophysics and astrophysics. Chapter 8 is a summary. The material comprising this subject has a very high degree of interconnectivity, as will become apparent, but I have found that its interrelatedness nevertheless allows a progressive presentation which leads to some interesting, if somewhat surprising, conclusions.

2. *Variable-G Cosmologies*

2.1. Introduction

Theories in which the Newtonian gravitational parameter, G, is time-variable or space-dependent are post-Einsteinian developments in the field of cosmology. It can be discerned in the six theories described below that the growth of interest in variable-G cosmology was a gradual process arising basically from the metric type of formalism, of which general relativity was the first notably successful instance. The sense of formal development in the subject was thus broadly chronological, and it is in this order that the following accounts are presented. Astrophysical aspects of those cosmologies are considered as they arise, as are geophysical ones that are not directly connected with the expanding Earth hypothesis. The latter is logically a separate concept from the variable-G hypothesis, since there are other theories that can explain the expansion, despite the fact that decreasing-G cosmologies actually predict such an effect. The expanding Earth is therefore deferred for discussion until Chapter 6, while this chapter gives a basic account of the modern cosmologies that go beyond the framework of general relativity.

2.2. Milne

Both the philosophy and the curved space–time of general relativity were abhorrent to Milne, who thought it an epistemological paradox to talk about space being curved. He built up a cosmology in flat space which is nowadays referred to as the Milne Universe, and which is set forth in his fundamental book *Relativity, Gravitation and World Structure* (Milne 1935). The starting point of Milne's approach is the Cosmological Principle, which is the statement that any two observers are

6

equivalent, in the sense defined by Milne (1935, p58). Two particle observers A and B are considered equivalent when A can describe the totality of his observations on B in the same way as B can describe the totality of his observations on A. The Cosmological Principle is the basis of all modern work on cosmology, and has been used in an extended form by Bondi and others with a postulate of temporal as well as spatial equivalence, when it is known as the Perfect Cosmological Principle.

Milne considered a population of particles in motion, and imposed his principle of equivalence on them. The resulting equation for the acceleration of a given particle is of the same form as the Newtonian acceleration at the same point, provided that $G \propto t$ (that is, the opposite dependence to that resulting from the majority of other non-general relativistic cosmologies). The $G \propto t$ dependence reproduces the effect of all the matter present, in so far as the latter is connected with local accelerations. To quote from Milne (1935, p103): 'The Newtonian constant G is of course defined to be independent of t. Our evaluation of an effective G, judged not from two ideal particles alone in space (as in the Newtonian definition), but from the actual local acceleration of a test particle in the presence of a certain distribution of particles in motion, gives G increasing secularly as the epoch measured from the natural zero of time'.

The appearance of Milne's cosmology was met with scepticism and puzzlement; it was difficult to see how Milne arrived at a system so dissimilar from that of general relativity, having started from the same declared principles. To elucidate this apparent contradiction, Robertson (1935, 1936a, b) and, soon afterwards, Walker (1935, 1936) examined Milne's approach closely, and both came to the conclusion that the world system of Milne was not the most general one possible. This might have been inferred directly from Milne's use of the Minkowski metric

$$ds^2 = c^2 \, dT^2 - [dr^2 + r^2 \, (d\theta^2 + \sin^2 \theta \, d\phi^2)] \qquad (2.1)$$

in comparison with the usual metric of general relativity, which contains the Minkowski form as a special case of the

7

more general metric

$$ds^2 = c^2 \, dt^2 - \frac{S^2(t) \, [dr^2 + r^2 \, (d\theta^2 + \sin^2 \theta \, d\phi^2)]}{(1 + kr^2/4)^2} \qquad (2.2)$$

where $S(t)$ is a function only of the epoch, and $k = +1, -1$ or 0. Robertson (1935) generalised Milne's fundamental transformation between observers, which is $t_1' = p_M(t_1)$, $t_2' = p_M^{-1}(t_2)$, taking t_1 and t_2 as two characteristic coordinates (similarly, t_1' and t_2'), to the more universal form: $t_1' = f(t_1, u)$, $t_2' = f(t_2, \bar{u})$, where u is a member of the one-parameter continuous group being considered ($\bar{u}$ is the inverse of u). Robertson thus showed that any space–time satisfying the Cosmological Principle admits, *ipso facto*, a metric of the form of equation (2.2). This metric is the most general one available that applies to a cosmology in which the matter density is homogeneous and isotropic. The model expands or contracts according to the behaviour of $S(t)$, which is a kind of scaling factor defining how the distance between two points varies as a function of epoch. The metric in equation (2.2) has the properties that (*a*) the world line of each fundamental observer is a geodesic of the metric ds^2, and the observer's clock time is measured along this geodesic; (*b*) the world line of every light signal is a null geodesic of the metric ds^2.

The above metric in equation (2.2), derived by Robertson, was also obtained by Walker (1936), and is now universally known by their joint names. As demonstrated by Robertson (1935), within the confines of the derivation there is no need to impose a particular law of gravitation. This was one of the advantages of Milne's approach—namely that it was not subject to any particular doctrine of gravity—but Robertson's work showed that no formalism consistent with the Cosmological Principle stated above can be wider in its scope than general relativity. At the time, Milne was developing his concept of what constitutes a class of equivalent particles (Einstein's general theory was not as firmly established as it later became). Milne laid more emphasis on the fact that a class of observers should be able to prove themselves kinematically equivalent using radar signals and clock regraduation than he did on a given theory of gravity or on a given set of field equations. The result was that he arrived at a theory that had an acceptable metric, but was not generally compatible with Einstein's theory. It also had the peculiar characteristic that two special time scales, τ and t, could be used to describe it with equal validity (see below). Both of these can be obtained from the general time coordinate, T, by a regraduation of clocks.

Milne's rigid formalism involving radar signals, based on the group of transformations given above, was made even more restrictive by the use of what he termed the 'dimensional hypothesis'. This amounted to insisting that certain key parameters in the theory, notably the one defining the nature of the particle distribution, should be dimensionless. This hypothesis also restricted the function $S(t)$. The result of the three principles involving equivalence, clock regraduation and dimensions was to define $S(t)$ and $S(\tau)$ in such a way that

$$S(t) \propto t; \qquad \tau \propto \ln t. \qquad (2.3)$$

There was thus a big bang in the t coordinate but not in τ, a property that has been repeatedly taken up over the years, and which finds its most complete usage in Dirac's cosmology (as discussed in §2.3). The question of which time scale related to which aspects of observed processes in physics and astronomy was not completely clear at the time Milne proposed his theory. It was apparent to him, however, that the difference between the two scales related to whether one was employing a small-scale atomic process or a large-scale dynamical process as standard, and that the difference in time scales could be expressed by making G in the second type of process depend on the time scale as employed in the first type of process.

Milne, in deriving $G \propto t$, did so by a method that requires G to be estimated from the acceleration of a free particle in space, and stated that this must not necessarily be taken to imply that local gravitation—in the Solar System, for example —possesses a varying G (Milne 1935, p292). This statement is based on the underlying tenet that the variability of G can only be discussed with regard to an atomic time scale, but has caused some confusion. To be consistent, if G were believed to be increasing (Milne calculated a rate of increase of one part in 2×10^9 yr^{-1}, but the data used are now believed to be greatly different), the Earth should be undergoing a con-

9

traction, in contradistinction to the expansion predicted by those cosmologies in which $G \propto t^{-1}$ (see §6.5). In former years, a contraction of the Earth was widely accepted as an explanation of mountain building. With advances in geophysics, the need for this contraction gradually vanished, although Jeffreys (1970, ch XI) remained as a notable proponent of it. His objection to expansion is that the original density of the Earth with the expansion hypothesis would have been greater than 20 g cm^{-3}, which he finds unacceptable. It could be argued that the density of the planet need never have been this high if continuous creation is operative (§6.4), as predicted, for example, by Dirac's complete theory. In any case, a density of about 20 g cm^{-3} is not greatly in excess of the present value of the density at the centre, which is estimated to be 16 g cm^{-3}. It is certainly important, however, that geophysical propositions, such as a varying G, should be kept within reasonable bounds.

Milne developed his theory along the lines of dynamics and electrodynamics, as opposed to kinematics, in a second book entitled *Kinematic Relativity* (Milne 1948). The contents of this sequel volume are now regarded as vitiated by certain mathematical errors contained therein, and Milne's world system is treated today as a particular case of the relativistic Robertson–Walker Universe models. It is an expanding, Minkowski-type system which is isotropic about every point, but in which the relativistic field equations are not imposed (Ellis 1971). It represents an explosion of matter into flat space–time, but in its original formulation, with Milne's attendant theory, it must now be considered obsolete, although certain innovations (such as clock regraduation) that were originated by Milne are still in widespread use in cosmology. In particular, the ideas of the Cosmological Principle, of an equivalent set of particles, and of the basis of kinematic relativity as it involves timed radar signals between points in an equivalent set of particles, are three concepts that were established by Milne, and which will be met with recurrently in the following account.

2.3. Dirac

The cosmology proposed by Dirac (1937, 1938) rests on the

principle that any two of the enormous numbers appearing in many theories of nature are connected by a simple mathematical relation in which the coefficients are of order of magnitude unity. Dirac rejected Milne's postulate that all cosmological numbers should be dimensionless. Instead, he selected on the ground of it being a mean value, e^2/m_ec^3 as a fundamental unit of time from among the following possibilities:

$$\frac{e^2}{m_ec^3}, \frac{e^2}{m_pc^3}, \frac{h}{m_ec^2}, \frac{h}{m_pc^2}, \frac{\hbar}{m_ec^2}, \frac{\hbar}{m_pc^2}, \qquad (2.4)$$

where e is the charge and m_e is the mass of an electron, m_p is the mass of a proton, c is the velocity of light, and h is Planck's constant with the dimensions mass $\times$ length^2 $\times$ time^{-1} ($\hbar \equiv h/2\pi$). The ratio of the present epoch of the Universe, T, to the unit of time e^2/m_ec^3 is approximately equal to 10^{39}. The ratio of electrostatic force to gravitational force between a proton and an electron is also $\gamma_D \approx 10^{39}$ ($\gamma_D = e^2/Gm_pm_e$ in CGS units, or $\gamma_D = e^2/4\pi \, G\epsilon_0 m_pm_e$ in rationalised MKS or SI units, where ϵ_0 is the permittivity of free space). The rough equality of these two large numbers suggested to Dirac that in fact $Tm_ec^3/e^2 = \gamma_D = e^2/Gm_pm_e$ holds as an *equality*. Since the epoch increases as some suitable time unit, t, the equality just given implies $G \propto t^{-1}$.

Assuming hc/e^2 and m_p/m_e to remain constant with epoch, and the validity of the conservation of matter, the density of the Universe must decrease as the expansion proceeds. In suitable units, the density of the Universe and the reciprocal of Hubble's parameter can be used with the principle again, to give relations of the form

$$\rho_U = k_D H = k \frac{\dot{f}(t)}{f(t)} \qquad (2.5)$$

where k_D is a constant (≈ 1) and f is a function describing how the distance between nebulae varies with time, t, in a Universe of mean density ρ_U. By making various additional plausible hypotheses, it transpires that $f(t) \propto t^{1/3}$, which leads to $t = f(t)/3\dot{f}(t)$. Consequently, the age of the Universe is only of order 10^9 yr (that is, 10^{39} units of e^2/m_ec^3). This is not an

11

objection to the theory, since the rate of radioactive decay must have varied in the past, and the data are open to interpretation with a more extended time scale.

This is the basis of the theory, but it has hidden ramifications since logically, as pointed out by Dirac, one might expect that all large numbers would vary with epoch. This rules out a closed Universe, since the latter is synonymous with a fixed total number of particles in the world. Similarly, simple hyperbolic space is connected with a constant mass within a given volume. There is left only a flat three-dimensional space, in which $G \propto t^{-1}$. The theory can be described in a different way by setting $G = $ constant and using a new system of units that varies in time with respect to the old system. This brings the hypothesis into conformity with conventional theory, in which G does not vary, and by taking $c = 1$, a new time–distance scale can be constructed. There now exist two time scales: one for atomic phenomena and one for mechanical, large-scale phenomena such as those usually treated in general relativity. The gravitational constant *is* constant in the latter, but G varies as t^{-1} in the former. The situation is thus very similar to that envisaged by Milne, but in the opposite sense.

Dirac proceeded further, making use of the Robertson–Walker metric, and putting the cosmological constant $\Lambda \equiv 0$ (since otherwise $\Lambda^{1/2}e^2/m_e c^3$ would be a dimensionless number and would have to vary with epoch, causing problems). It is interesting to note that Dirac mentioned continuous creation of matter in proposing his new theory, but disregarded it. The main contention of Dirac's argument was that G could be variable. This is not in conformity with Einstein's theory, since it is generally accepted that general relativity, unlike Dirac's cosmology, admits no possibility of G being variable. The reason is that G is simply a conversion factor that enters because mass is usually measured in other than gravitational units (see Chapter 3). The same applies to the velocity of light (McCrea 1971). The possible variation of G suggests that other fundamental parameters of physics may be variable. This subject is connected closely with that of the Eddington numbers (of which those employed by Dirac of size 10^{40} are examples), and these are discussed fully in Chapter 4. Dirac's original hypothesis was essentially an attempt to link together some of the singularly large numbers of astronomy by dimensional analysis. It lacked a deeper theoretical basis and it was with the object of providing this that Jordan took up the subject as described below.

Many years after Dirac proposed his large-numbers hypothesis in the familiar $G \propto t^{-1}$ form, he completed the theory by adding to it a consistent cosmological basis involving particle creation (Dirac 1973a, b, 1974). The original theory in which $G \propto t^{-1}$ and e^2/Gm_em_p ($\approx 10^{39}$) $\propto t$ with numbers of order $(10^{39})^n \propto t^n$ and the numbers of particles in the Universe proportional to t^2 (Davies (1974d) has revised this form of the hypothesis), has now been extended by employing two metrics, one relevant to general relativity and one to laboratory (atomic) physics. The large-numbers hypothesis is essentially incorporated by allowing the metric intervals in the Einsteinian and atomic regimes to be unequal, ds (Einstein) $\neq ds$ (atomic), where the ratio of the two (ds_E/ds_A) is a function of time that can be expressed as a dependency of G on time. Such an approach is similar to that of the old Weyl theory (see Weyl 1922, Dirac 1973a, b, c, 1974) in which electromagnetism and gravity are unified, but in which the length of a vector is non-integrable when carried around a closed loop in the field. The cosmological background, because it is non-integrable, needs to be referred to a metric gauge, but ds_A does not, and is unaffected by such transformations of the potentials (the g_{ij} in $ds^2 = g_{ij} dx^i dx^j$). The field equations for the theory are obtained from an action which is an integral over a co-scalar Ω_D, which is itself a combination of the usual electromagnetic action and gravity (see below). The resulting formalism, which is similar to that of the Brans–Dicke scalar–tensor theory, is in a sense superior to general relativity in that it makes physics invariant under transformations of curvilinear coordinates in Riemannian space (Einstein) *and* transformations of gauge (the latter being basically the choice of an arbitrary standard of length at each point). This type of theory leads to a breaking of C and T invariance, but not of P or CT, and the consequent long-range forces and broken symmetries lead to the cosmological effects considered in §2.8 (see Dirac 1973b for his account of broken symmetries). It also leads, because of the relation involving the number of particles in the Universe noted above, to continuous creation.

The creation of new atoms (Dirac 1974) can take place in two forms: additive or (+) creation is creation at a uniform rate everywhere, while multiplicative or (×) creation is creation where matter is already most dense. The two forms of creation can be differentiated by the ways in which the atomic metric, ds_A, varies as a function of the Einstein metric, ds_E, the comparison between the two being most conveniently made at the Solar System scale. If the Earth orbits the Sun (mass $M_\odot$) at a distance, r_E, with velocity, v_E, in Einstein units, then the usual Newtonian relation gives $G_E M_{\odot E} = v_E^2 r_E$. The corresponding relation in atomic units is $G_A M_{\odot A} = v_A^2 r_A$. In Einstein units, G_E, $M_{\odot E}$, v_E and r_E are all constants. In atomic units, $G_A \propto t^{-1}$, and depending on the type of creation operating, either $M_{\odot A} =$ constant (additive, number of nucleons = constant) or $M_{\odot A} \propto t^2$ (multiplicative, number of nucleons $\propto t^2$). Thus $r_A \propto t^{-1}$ for (+) creation, $r_A \propto t^{+1}$ for (×) creation, since v_E expressed in units of the velocity of light, c, is dimensionless and is a constant. For the metrics, $ds_A \propto t^{-1} ds_E$ with (+) creation, or $ds_A = t\, ds_E$ with (×) creation. With (+) creation, which is supposed to occur mainly in intergalactic space, the Solar System contracts (in atomic units), while with (×) creation, which is supposed to occur inside astrophysical bodies, it expands. It is informative to examine the (+) and (×) types of creation in slightly greater detail.

Additive creation, as noted above, entails $ds_A \propto t^{-1} ds_E$, so that putting $ds_A = c\, dt$ allows one to define the Einstein epoch as

$$\tau \propto \int ds_E \propto \int t\, dt = t^2/2.$$

This dependency means that the cosmology is one of infinite extent with hyperbolic curvature. There was a big bang origin to the model, but since the masses of large bodies in Einstein units are a constant for (+) creation, as are the masses of nucleons, the overall density of matter decreases as τ^{-2} as the model expands. The created matter must be equally comprised of positive and negative mass in order that the cosmology obeys the principle of conservation of matter built into its equations.

14

It is assumed that the positive mass condenses to form stars, galaxies, etc, while the negative mass remains as a spread-out, unobservable background. The $(+)$ creation cosmology is thus similar to the C-field theory of Hoyle which will be commented on below.

Multiplicative creation gives an entirely different cosmology, which arises by allowing a cosmological constant to enter the equations. The latter, apart from the symmetry-breaking term, are essentially the same as Einstein's equations, so it is apparent that, if Λ is taken to have dimensions of time^{-2}, $c\Lambda^{-1/2}$ is a cosmological length. If this length is divided by some atom-sized distance, δs_E, it gives a dimensionless number, and $c\Lambda^{-1/2}/\delta s_E$ is, in fact, of the order of 10^{40}. If one assumes that this number increases as t by the large-numbers hypothesis, one finds $\delta s_E \propto t^{-1}$, since Λ does not vary, whereas δs_A is constant (atomic units). This means that $(\times)$ creation is involved. The finiteness of Λ and $(\times)$ creation are indisputably linked, whereas $(+)$ creation can only occur if $\Lambda \equiv 0$. The $(\times)$ model thus differs in principle from Dirac's original theory, which took Λ to be zero. When $(\times)$ creation is involved, atomic units as used in laboratory and smaller-sized systems have a corresponding dynamical time appropriate to larger systems, which is

$$\tau \propto \int \mathrm{d}s_E \propto \int t^{-1}\,\mathrm{d}t = \ln t.$$

The big bang occurred at $t=0$ or $\tau = -\infty$. G is a constant on the τ scale. This is similar to Milne, and effectively means that no evidence of a big bang should exist on the large scale at the present epoch. While parameters are constant on the large scale in terms of Einstein units, atomic parameters are not. It is necessary to choose an atomic unit of mass that varies as t^{-2} in order to cancel the corresponding increase due to $(\times)$ creation in the particle number of a large body, and so ensure that the mass of a body such as the Sun is a constant as expressed in Einstein units. Thus the mass of a nucleon in Einstein units is $m_n \propto t^{-2}$. The relation $e^2/Gm_n^2 \propto t$ of the large-numbers hypothesis in its original form still holds, and it can be used to see how atomic parameters vary when referred to Einstein units. In these units, G is a constant, $m_n \propto t^{-2}$, so that $e \propto t^{-3/2}$. These dependencies result in Planck's constant being $h \propto t^{-3}$.

The strengths of these dependencies on atomic time t (which is approximately synonymous with the practical Atomic Time service mentioned elsewhere), depend on the time that has elapsed since the big bang ($t=0$, $\tau = -\infty$). The dependencies of atomic parameters on Einstein units and the dependencies of large-scale systems on atomic units severely affect cosmology. All expanding models have an epoch which characterises their evolution, but the epoch as measured in atomic units ($e^2/m_e c^3$) is a *constant* large number of the order of 10^{40}. This is not in agreement with the postulate that all such numbers should vary as powers of t in the manner noted, and therefore only static models are allowed, in agreement with the comment made above. In particular, the Einstein static Universe is taken to be the basic cosmology. The cosmological constant Λ is vital, since the static Einstein Universe is kept distended only by the Λ force with $\Lambda > 0$ and $G \equiv$ constant. If Λ is finite, therefore, ($\times$) creation is necessarily operative if the new Dirac theory holds.

The ($\times$) theory accounts for the redshifts of the galaxies in the following way. While an atomic clock ticks off units $\Delta t = 1$, these correspond to units of $\Delta \tau \propto t^{-1} \Delta t$. An atomic clock was therefore slower running in the past when referred to Einstein units. In these units, the wavelength remains unaltered during transmission, but on arrival is compared with a clock (that is, an atom) that is running faster than the one at the place where it was emitted. The result is that a redshift is observed which is $t(\text{now})/t(\text{emission}) - 1$.

The way in which ($\times$) creation differs from ($+$) creation provides several opportunities to carry out observations designed to test the Dirac theory and continuous creation in general.

The additive and multiplicative models have been tested by Chin and Stothers (1975), who ran stellar evolution programs suitably modified for both possibilities (the effect of decreasing G causes the luminosity of a star to decrease, while an increasing mass causes its luminosity to increase; see below). The familiar Pochoda and Schwarzschild calculations (§2.6) have, in fact, tested the ($+$) model, and, while it is found to be acceptable if the age of the Universe is $T \gtrsim 1 \cdot 5 \times 10^{10}$ yr, the new calculations of Chin and Stothers would seem to rule it out by comparison with the Sun's luminosity ($L_\odot$) and age. On the other hand, the ($\times$) model is acceptable, since the ancient

16

higher value of G, but smaller value of $M_\odot$, would have approximately cancelled ($M_\odot \propto t^2$, $G \propto t^{-1}$), so that the ratio $L_\odot/M_\odot$ has been roughly constant if the ($\times$) process has been operative. The temperature of the Earth's surface has varied only slowly, as $(L_\odot/r_E^2)^{1/4}$, due to the secular increase in the radius of the Earth's orbit as t^{+1}. Chin and Stothers remark that since the upper mass limit for stable white dwarfs or neutron stars is proportional to $G^{-3/2} \propto t^{3/2}$, with $M_* \propto t^2$, it is expected on the ($\times$) model that all astrophysical objects will eventually end up as collapsed bodies or black holes.

The doubt raised by Dirac (1974) concerning the ($\times$) process and the distortion of ancient rocks has been answered by Gittus (1975), who has shown that the creation of new atoms in old rocks would not cause disruptions, because the crystal structure would already possess imperfections, sinks and dislocations into which the newly created atoms could diffuse. Gittus suggested that ($\times$) creation could be tested by looking for the expected changes in the density of dislocations over time in, for example, Moon rocks, since dislocations should lengthen if atoms are being continually created in the multiplicative manner. Gittus' view of the admissibility of ($\times$) creation has been criticised by Towe (1975), who remarks that while 30% of new atoms could appear over a rock age of 3×10^9 yr without disruption, ($\times$) creation is wrong, as is ($+$) creation, unless Avogadro's number is constant in time. He arrives at this from the formula $V = M_w n_a / \rho_x N_A$ of x-ray crystallography, where V is the lattice volume, M_w is the molecular weight, ρ_x is the x-ray density of the crystal unit cell, n_a is the number of atoms and N_A is Avogadro's number. The parameters appearing in this equation are observed to be the same at the present epoch as they were in the Pre-Cambrian, which he finds to be incompatible with a variation involving $n_a \propto t^2$. In turn, Towe has been criticised by Canuto et al (1976) who point out that, in practice, no effects of the type considered could be detected in crystals because the instruments used in crystallography all effectively use atomic time, as do the constituents of the crystal being examined. Towe (1976) agrees with this criticism, but considers that since masses grow as t^2 (mass $= N_A m_a$ for atoms of mass m_a), the expected 30% of new atoms should still have been found if ($\times$) creation is operative.

17

He believes that the only way out of this dilemma is to have the volume of the crystal growing as t^2 as well.

One possible other alternative has been proposed by Steigman (1976), who has sought to demonstrate that the large-numbers hypothesis of the original Dirac theory predicts that the number of atoms in the Universe grows as t^2 without the need for the newer Dirac mechanism of continuous creation. Steigman's contention rests on the fact that the horizon of the Universe is growing continuously as some function of cosmic time, t. While the density decreases, the volume can increase faster, so that the net number of particles in the Universe grows as some power of t, this power depending on the exact model. Steigman is thus able to choose a model with a scale factor, $S(t)$, that gives a particle number within the horizon increasing as t^2. His argument, however, has been criticised (Adams et al 1976). If $S(t) \propto t$, as in one possible Friedmann-type model, then there is no horizon, and the idea fails (although it might still work if $S(t) \propto t^{1/3}$, since in that case there is a horizon). Steigman's interpretation does not therefore appear to be equivalent to Dirac's in general, but it is useful to consider its consequences for the microwave radiation field and nucleosynthesis in the early Universe, since these are so unacceptable that they seriously compromise the original large-numbers hypothesis of Dirac.

It would therefore seem that the new (as opposed to the old) formulation of Dirac's hypothesis must be taken, if any agreement with astrophysical data is to be obtained. Apart from the crystal problem, the new theory implies that the Earth's mean temperature was different in the past because of the evolution of the planet's orbit around the Sun (Roxburgh 1976). The continuous creation aspect of the new formulation has implications for all rotating systems in astrophysics, since the theory predicts that angular momenta decay secularly. This has been used by Mansfield (1976), in a study of pulsar spin-down, to show that the original Dirac theory, the Hoyle–Narlikar theory and the new Dirac ($+$) creation model are probably unacceptable, but that the ($\times$) creation model does not conflict with the spin-down rate of the pulsar JP1953. The quasi-Newtonian effects of ($+$) creation and ($\times$) creation on stellar bodies have also been considered by Stothers (1976). He discusses the stability of white dwarfs via the effects of secular changes in the particle number and the Chandrasekhar limiting mass (the latter being $\propto G^{-3/2}$). For $(+)$ creation white dwarfs become more stable with time, whilst with $(\times)$ creation the older stars tend to collapse to black holes or neutron stars. Available astrophysical data on the effects of these contending influences tend to support $(\times)$ creation as more acceptable. A similar conclusion has been reached by Faulkner (1976), who has studied the effect on the acceleration of the Moon's mean longitude of a decrease in G at the rate suggested by Van Flandern (see below), and has discussed the conservation of angular momentum and related topics as they affect astronomical bodies.

It is normally acceptable, as an approximation, to replace the usual Newtonian gravity law by one in which $G = G(t)$ in systems of stellar size. Bishop and Landsberg (1976) have shown, however, that it is *not* automatically valid in systems of large size, unless other effects are taken into account. In a system with a very large number of particles, for example, simply replacing G with $G(t)$ will usually lead to a violation of conservation of energy for the whole system. A consistent variable-G scheme in the Newtonian regime can be worked out, but it involves Machian-type forces. These depend on the relative velocities of the particles in the ensemble, with an analogue of the zero-point energy of quantum physics (that manifests itself in the impossibility of any two particles being at relative rest to each other). The force of gravity in such a multi-component Newtonian system has to be written as a term of the usual sort, with $G = G(t)$, plus a second term that incorporates the new effects but is of comparatively minor influence in most astrophysical problems. Clearly, however, one must be careful to appreciate all possible consequences, for a given problem, of applying both the variable-G and continuous creation concepts to astrophysical systems in the quasi-Newtonian limit.

2.4. Jordan

Jordan, in an important paper (Jordan 1949), realised that if A_* denotes the age of the oldest stars ($\approx 10^{17}$ s), H is Hubble's parameter, ρ_U is the mean mass density of the Universe and R_U is the radius of the Universe, then the numbers

$$HA_*, \qquad R_U/cA_*, \qquad \gamma'\rho_U c^2 A_*{}^2 \qquad (2.6)$$

are all dimensionless and of order unity ($\gamma' \equiv 8\pi G/c^2$). Assuming that these numbers are precisely equal to unity gives $HA_* = 1$, $R_U = cA_*$ and $\gamma'\rho_U c^2 A_*{}^2 = 1$. Thus, the radius of the Universe increases at the velocity of light, and from the last precise equality, putting the mass of the Universe as $M_U = \rho_U R_U{}^3$, it is seen that

$$\gamma'(M_U/R_U{}^3)c^2(R_U{}^2/c^2) = 1, \qquad \text{or} \qquad \gamma' M_U = R_U. \quad (2.7)$$

That is,

$$M_U G = c^2 R_U/8\pi. \qquad (2.8)$$

Since R_U, the radius of the Universe, is changing, the other parameters also change, and the proportionality $G \propto t^{-1}$ results. In this cosmology, it is possible for the total energy of the Universe to be zero, since the rest mass energy $M_U c^2$ can be made equal to the (negative) gravitational potential energy of the system. (This occurs because equation (2.8) is equivalent to $GM_U{}^2/R_U \approx M_U c^2$.) Jordan proposed that stars are formed as whole bodies, in contradistinction to all other theories of stellar formation, but in conformity with the law of conservation of energy. He has applied his theory (Jordan 1969) to the expanding Earth hypothesis (discussed in Chapter 6) and to astrophysical situations affected by the $G \propto t^{-1}$ dependency.

Jordan, by taking the Sun's luminosity to vary as $M_\odot{}^{11/2} R_\odot{}^{-1/2} G^{15/2}$, deduced that the solar constant varies as G^{10}. This is a very strong dependence on G, and means that the surface temperature of the Earth during the Carboniferous would have been much hotter than at present, conditions being somewhat like those prevailing on Venus at the present epoch. This is obviously hardly acceptable. Hoyle and Narlikar (§2.7) derive much less extreme conditions. Jordan's theory, although dubious from the geophysical aspect, was important in providing (Jordan 1962a, p597) a theoretical basis for Dirac's hypothesis by the use of an action principle.

To give a theoretical basis to geophysical and astrophysical speculations which followed the manner of Dirac's previously proposed hypothesis, Jordan sought to generalise Einstein's general theory of relativity by changing the variational principle from which the latter can be derived. He replaced the variation

$$\delta \int \left(R_S - \frac{\gamma' F_{kl} F^{kl}}{2c^2} - \frac{\gamma' L}{c^2} \right) dV^4 = 0, \qquad (2.9)$$

which is written using the notation of Jordan (1962a) for the usual components of the action (Synge 1966), by the modified form

$$\delta \int \left(\frac{R_S}{\gamma'} - \frac{\xi_J \gamma'^{,j} \gamma_{,j}'}{\gamma'^3} - \frac{F_{kl} F^{kl}}{2c^2} - \frac{L}{c^2} \right) dV^4 = 0. \qquad (2.10)$$

Here, $dV^4 = (-g)^{1/2} dx_1 dx_2 dx_3 dx_4$, $\gamma' \equiv 8\pi G/c^2$ is the (now variable) relativistic gravitational parameter, and ξ_J is a dimensionless constant (≈ 10). $\gamma_{,j}'$ denotes the ordinary gradient of the scalar γ', and R_S is the scalar curvature. From equation (2.10), Jordan (1962a) obtained three Lagrange equations, one of which defines F^{kl} by $F_{;l}^{kl} = 0$, where the semicolon denotes the covariant derivative.

The theory has been greatly developed by Jordan (1955, 1962b, 1964), notably in investigating the implications of Dirac's hypothesis for geophysics (as influenced by $G \propto t^{-1}$) and cosmogony. In the latter context, he has shown (Jordan 1967) that the concept of Ephemeris Time (that is, the time scale given by the motion of the planets) remains definable, but the time scale so derived is not equal to the time coordinate of any inertial system. Several of Jordan's co-workers have studied this cosmology. Trümper (1963) has obtained the exterior solutions to the scalar χ_J field of the theory, which is similar to the scalar–tensor theory of Brans and Dicke discussed in §2.6. Unfortunately, the χ_J theory was to a large extent vitiated when Dehner and Hönl pointed out that the then newly discovered 3 K background radiation implied that the photon density and spectrum of the background had not altered over the lifetime of the Universe. This result was irreconcilable with the particle creation interpretation of Dirac's hypothesis by Jordan (Dehner and Hönl 1969, Jordan 1968). The discarding of Jordan's original field equations and the subsequent adoption of new ones (Jordan 1968) made the Jordan cosmology into a very close parallel of the scalar–tensor theory of Brans and

Dicke. The resulting non-general relativistic cosmology is frequently referred to by a combination of the three names.

2.5. Gilbert

It was mentioned above that a finite rate of change of G with time $(\dot{G}\neq 0)$ has not generally been considered to be consistent with general relativity. In contradistinction to other workers, Gilbert (1956a) has claimed that the amalgamation of Dirac's hypothesis with general relativity is feasible, and states that general relativity is not inconsistent with a finite value for $\dot{G}$. Gilbert's procedure consisted in taking two sets of units, one set to be used in describing electromagnetic phenomena, the other set to describe gravitational phenomena. In one set, the Newtonian law of gravitation is not a function of time, and cosmic distances are employed. The other set is used with a Newtonian law of gravitation in which the gravitational power of matter decreases with time as t^{-1}. The hypothesis is valid so far, since electromagnetism is independent of Newtonian gravitation, and there is no reason why the coordinates used in the two areas should be the same. The analysis has something in common with the theories of Milne (who, however, used kinematic relativity) and of Jordan (who used projective geometry). The ratio of the electromagnetic system and the gravitational system in Gilbert's account varies with time, of course, and he makes the theory more explicit by postulating the following:

(i) An atomic system in which the metric is immediately specified as the familiar

$$ds^2 = c^2\,dt^2 - S^2\,(dx_1{}^2 + dx_2{}^2 + dx_3{}^2) \qquad (2.11)$$

where $S = S(t)$ only, and the 3-spaces $t =$ constant are conformal to Euclidean space, in which $r = (x_1{}^2 + x_2{}^2 + x_3{}^2)^{1/2}$. The coordinates t, x_1, x_2 and x_3 are called the basic coordinates; they are the often-quoted comoving coordinates in which the coordinate distance between any two particles is constant. In this system, Gilbert derives $S \propto t^{2/3}$, and postulates that all species of fundamental particles (for example, protons and electrons) have associated with themselves intrinsic units of ϵ s and δ cm.

(ii) The cosmic coordinates mentioned previously are now introduced by defining $X_\alpha = S x_\alpha$ ($\alpha = 1$, 2 or 3), where the cosmic coordinates (t, X_1, X_2 and X_3) are to be used by an observer in Euclidean space. Maxwell's equations have their natural description in terms of cosmic coordinates. An observer will discover a closely Newtonian law of gravitation in which the usual (general relativistic) gravitational coupling parameter is found to be given by $8\pi\Gamma/c^2$, Γ being analogous to G. The derivation of this involves an equation for the mass contained within a sphere of given radius. The observer arrives at the Newtonian gravitational law, provided he assumes that matter beyond the radius of this sphere gives no resultant gravitational force on matter inside it (following Milne).

When cosmic coordinates are employed (t, X_1, X_2 and X_3), the gravitational field of a static mass is the Schwarzschild solution

$$\mathrm{d}s^2 = c^2 \left(1 - \frac{2\Gamma M}{c^2 r'}\right) \mathrm{d}\tau^2 - r'^2 \,(\mathrm{d}\theta^2 + \sin^2\theta\, \mathrm{d}\phi^2) - \frac{\mathrm{d}r'^2}{[1 - (2\Gamma M/c^2 r')]}$$

$$(2.12)$$

where the usual nomenclature is employed and $\Lambda \equiv 0$. The time τ appearing in the Schwarzschild metric is a function of the cosmic time, t. When velocities are small compared with the velocity of light, and $GM/c^2 r' \ll 1$, the interval $\mathrm{d}\tau$ is approximately equal to $\mathrm{d}s/c$. The propagation of light in a vacuum is in agreement with Maxwell's equations, provided cosmic time and space coordinates are used in them. On the other hand, the relativistic gravitational equation can be put into Newtonian form if non-cosmic coordinates are used. In the Newtonian system so specified, G is a function of time. Let the non-cosmic coordinates be $\zeta_\alpha = \lambda_G x_\alpha$, where $\lambda_G = \lambda_G(t)$ is slowly varying, then Gilbert has shown that $\lambda_G = C_G^{(1)} t^{2/3} + C_G^{(2)} t^{1/3}$, where $C_G^{(1)}$ and $C_G^{(2)}$ are constants. The cosmic coordinates are recovered if $C_G^{(2)} = 0$ and $C_G^{(1)}$ is a function of the constants of physics having connection with the units ϵ s and δ cm introduced above. The case $C_G^{(2)} = 1$, $C_G^{(1)} = 0$ gives a third system of so-called local coordinates, in which $G \propto t^{-1}$ results by comparison of the Newtonian law of motion involving G

23

with the relativistic system involving Γ. This demonstration of $G \propto t^{-1}$ is basically a result arrived at by assuming that a local observer tries to use cosmic coordinates to measure Newtonian gravitational phenomena. Note, however, that if basic (comoving) coordinates were to be introduced, $S \propto t^{2/3}$ would make λ_G a constant and so G a constant. The variability of G is therefore primarily a question of choice of time parametrisation, as realised by Milne (§3.8). If an observer uses cosmic coordinates to describe electromagnetic phenomena and local coordinates for gravitational phenomena, dimensional analysis involving ϵ, δ and the proton mass, electron mass, velocity of light and G ($\propto t^{-1}$) can be shown (Gilbert 1956a, p689) to provide estimates of the age of the Universe, Hubble's parameter, H, and the mean universal matter density. All are within an order of magnitude of agreement with observation.

Gilbert's main contention, that all quantities occurring in field theory should be based on mass, length and time units derived from the values of the physical parameters measured in the local frame, would seem to be a reasonable postulate. Gilbert has remarked (personal communication) that while earlier work (Gilbert 1956a) on the interpretation of a time-variable G in cosmology has relevance to both relativistic and Newtonian systems, yet what was really needed was a conformally invariant theory (Gilbert 1960; see §3.9 for an account of conformal invariance). However, the latter attempt was not entirely successful, although some of the conclusions arrived at were similar to those since found by Hoyle and Narlikar from the direct-particle interaction theory.

The Gilbert (1960) theory is interesting on account of the clear physical way in which G variability enters. Einstein's equations can be derived from the principle of stationary action with an action density, $R_S (-g)^{1/2}$, as in equation (2.9), where R_S is the scalar curvature and g is the determinant of the metric tensor, $g = |g_{ij}|$. The linear presence of R_S means that the action density is identically zero, so that matter must be situated at singular events. Gilbert used the quadratic action principle

$$\delta \int R_S^2 (-g)^{1/2} \, dV^4 = 0,$$

where dV^4 is an element of coordinate 4-volume, to derive

24

field equations in which the field of the matter can be determined without needing to make use of the energy-momentum tensor T_{ij} explicitly (since only the change in R_S is involved). The field equations derived have affinities with the unified electromagnetic and gravitational theory of Weyl (1922) and that of Lanczos (1957), which are effectively the same as those of Dirac's new theory discussed above. Gilbert develops the theory by assuming the Cosmological Principle and Mach's Principle; the latter is required to make paths continuous at matter–vacuum interfaces (see Chapter 7). He also takes as a postulate that the action density is invariant under transformations to conformal space–times. The average, large-scale distribution of matter defines the natural gauge, while the locally centrally symmetric field around any individual mass defines the local gauge. It is assumed that the natural gauge gives a fundamental system of measurement in atomic physics to agree with Dirac's hypothesis. Inertial mass is defined from the form of the action for a massive particle; it is found that the mass, m, varies as $m \propto T^{1/2}$, where T is the epoch in the inertial system in which a free particle has constant momentum. Near a massive particle, Newton's equations hold approximately, but with the gravitational power of matter varying as $T^{-1/2}$. Gilbert states that the different scales arise because the discontinuity in the action at a three-space separating differing regions of space–time implies that some of the fundamental tensors and their first derivatives may be discontinuous there. This is not, strictly speaking, correct (cf Chapter 7), but the difference in scales may also be justifiable on the basis that there is no *a priori* reason why the Riemannian geometry of different regions should be based on the same standards of internal measurement, although suitable gauge transformations can enable the geometry of the whole space–time to be described in terms of one standard.

Gilbert's theory has cosmological solutions of the Robertson–Walker type, familiar from general relativity, with varying behaviours depending on the value of the cosmological constant. In general, however, the incorporation of Dirac-type dependencies into general relativity, even as allowed by reparametrisation, is not consistent because the two scales involved are discrepant and the change in units would make the electronic charge variable with time, as pointed out by Bondi (see Gilbert 1960, p378). This is contrary to the principle of conservation of charge inherent in Einstein's theory. Whereas the local gauge has the same metric as the Einstein–de Sitter Universe, the physical Universe must be measured by the gauge defined by the gravitational field of all the matter, if Dirac's ideas and Mach's Principle are to be correctly incorporated. Milne's approach to similar problems is also included to a certain extent in Gilbert's work, since the $T^{-1/2}$ dependency gives orbits around a central particle that are equiangular spirals (as was first calculated by Jeans). Milne realised that matter that becomes detached from a centrally rotating nucleus of a galaxy will, if it moves in an equiangular spiral, result in the formation of a spiral arm (Milne 1946, 1948). Results such as this do not necessarily require any departure from general relativity.

Gilbert (1956b) used Einstein's theory to calculate the effect of the expansion of the Universe on a globular cluster of stars embedded in it, showing that the cluster is stable provided its size is less than 0·4 times the size of a scale, S_G, provided by the cosmology. The scale is given by an equation for the scale factor, $S(t)$, that is $(dS/dt)^2 = K_G/S - c^2/S_G^2 + \Lambda c^2 S^2/3$, where K_G is a constant. The size of S_G therefore depends on the cosmology, and may not even be real (Gilbert 1956b, p682). This implies that in some cosmologies the expansion of the Universe can result in cluster configurations that are not necessarily secularly stable. This means, in turn, that motion on a time scale of the evolutionary age of the cosmology is a different kind of phenomenon from motions on shorter time scales and smaller scales (see Chapter 7). The basic fact would appear to be that expanding universes can, under a range of conditions, result in space–time dilation in any local domain that affects the metrics of clusters in a finite way. This phenomenon is small, but calculable, and results in non-Newtonian (although Einsteinian) processes in the evolution of clusters and other astrophysical systems. Although there are more conventional explanations, spiral arm formation and other problems in astrophysics could be profitably investigated using the wider non-Einsteinian formalism of Gilbert and others. This might provide a way of discriminating between his theory and the more fully developed Brans–Dicke and Hoyle–Narlikar cosmologies.

2.6. Brans and Dicke

The cosmology developed by Brans and Dicke (1961) is the most extensively investigated of those modern theories in which G depends on time. The relativity formulations of Brans–Dicke and Hoyle–Narlikar predict finite values for $\dot{G}/G$. These can be compared with the limit of $\dot{G}/G \leqslant 4 \times 10^{-10}$ yr^{-1} set by Shapiro *et al* (1971) using a method of monitoring the planets for a possible secular increase in their orbital periods, a radar reflection system being employed between the Earth, Venus and Mercury. The limit on $\dot{G}/G$ set by Shapiro *et al* is ten times greater than the change predicted by Dicke from the Brans–Dicke scalar–tensor theory. The theory is therefore acceptable, in the sense that it does not contradict this limit. It should be noted, however, that as more data accumulate the limits on the admissibility of variable-G theories become more stringent, and with regard to the Brans–Dicke theory (Shapiro *et al* 1976), the limits are particularly restrictive, as will be seen below.

The scalar–tensor theory grew out of the partial failure of general relativity to conform to Mach's Principle and the wish to conform with Berkeley's criticism of Newtonian space. In addition to vector fields (such as electromagnetism) and tensor fields (such as gravity), an additional scalar field is postulated which, however, successfully masquerades as gravitation. Consequently, experiments such as the bending of light around the Sun do not readily distinguish between the general relativistic and Brans–Dicke formulations (Hill 1971). The theory has affinities with Jordan's hypothesis, in that the units for mass, length and time are chosen from one of the two sets

$$(hc/G)^{1/2}, \ (Gh/c^3)^{1/2}, \ (Gh/c^5)^{1/2} \qquad \text{or} \qquad m, \ h/mc, \ h/mc^2$$

$$(2.13)$$

where m is the mass of an elementary particle. In choosing the second set, particle masses become constant by definition, but G is unavoidably made a variable: $G = G_0 \lambda_{\mathrm{BD}}^{-1}$, where G_0 is a constant and λ_{BD} is a dimensionless scalar field factor. Jordan was the first to formulate the equations of the theory along these lines, and the Brans–Dicke theory emerged from these equations with the added influence of Mach's Principle (Brans and Dicke 1961). Mathematically, the derivation of the theory is analogous to the method adopted by Jordan, using a modified variational principle (Dicke and Peebles 1965). As seen in §§2.4 and 2.5, general relativity can be deduced by a variation according to

$$\delta \int L_A \, d\eta = 0,$$

where η is an arbitrary parameter characterising a curve in space, $x^i = x^i(\eta)$. The L_A used in general relativity is of the form

$$L_A = eA_iU^i + m(g_{ij}U^iU^j)^{1/2} + q(h_{ijk}U^iU^jU^k)^{1/3} + \ldots \quad (2.14)$$

assuming space to contain things such as a covariant vector field $A_i(x)$, a symmetric tensor $g_{ij}(x)$, and $h_{ijk}(x)$, etc (e, m, q, etc are constants; $U^i = du^i/d\eta$). The $A_i(x)$ term is connected with electromagnetic phenomena, the $g_{ij}(x)$ with gravitational phenomena, while the h_{ijk} and higher terms in equation (2.14) are not observed and are logically undesirable. The Euler equations obtained from equation (2.14) give the familiar Lagrange equations, and these give the usual relativistic equations of motion.

The L_A of equation (2.14) has no scalar term. This could be introduced by using a more complicated L_A involving the products of fields, some of the terms in which, when multiplied by scalar field variables, could be absorbed into the vector or tensor part by simply redefining the latter. An equivalent procedure is to replace the second term of equation (2.14) with a new term of the form

$$m\phi_{BD}(x)(g_{ij}U^iU^j)^{1/2}. \quad (2.15)$$

The Brans–Dicke theory emerges from this step with the redefinition of g_{ij} to $\bar{g}_{ij} = g_{ij}\phi_{BD}^{-2}$ (where ϕ_{BD} is a scalar function of position). By this redefinition, the scalar field formally disappears from the equations of motion for a particle. But it is observable, since the wave equation for A_i or g_{ij} is different, depending on whether or not a scalar field has been absorbed in this manner. The theory can be formulated either in a way that makes $G = G(t)$, or in one in which the dimension-

28

less gravitational coupling constant $Gm^2/\hbar c$ ($\approx 10^{-40}$ by laboratory measurement) is a function of ϕ_{BD}. This arises since $m = m_p$ (the mass of the proton) is a function of ϕ_{BD} in the Brans–Dicke derivation. The theory is well known today in the form in which it predicts $G = G_0(t/T_0)^{-\gamma_{BD}}$, $\gamma_{BD} = 2/(4 + 3\omega_{BD})$, where ω_{BD} is a parameter that can be used to obtain agreement with observational data and specifies the strength of the coupling. It is usually the differences between general relativity and the scalar–tensor predictions of gravitational effects that are examined with regard to testing the Brans–Dicke cosmology.

If a part (perhaps 5%) of the agency that we know as gravity is due to a scalar interaction, the advance of the perihelion of Mercury should be that much less than the value predicted by general relativity. The observed advance is interpreted by Dicke to require the parameter ω_{BD} of the theory to be such that $\omega_{BD} \simeq 6$. McVittie (1969) has pointed out that a spherically symmetric Sun and the observed perihelion motion indicate $\omega_{BD} \geqslant 30$, which is not substantially different from Einstein's theory. It is well known that the observed advance of the perihelion agrees so well with Einstein's prediction that the scalar–tensor theory requires the somewhat *ad hoc* addition of a certain oblateness of the Sun to account for the observations. This is a weakness of the theory (Roxburgh 1969), as is the doubt that the theory's prediction agrees with the observed path of a light ray in its passage near the limb of the Sun.

The perihelion advance of Mercury, while in close agreement with general relativity, would not be so if part of the advance was due to an excess oblateness of the Sun. Dicke has consistently advocated this claim in support of the scalar–tensor theory (for example, Dicke 1970). His belief in the existence of an excess oblateness that would adversely affect the acceptability of Einstein's theory seemed at one stage to have been vindicated, since an attempt to confirm the excess was positive (see Davies 1974b). However, this result has been much debated, since the reality of the departure needed by Dicke depends for its interpretation on various astrophysical processes connected with the solar rotation, temperature and density gradient. In particular, Hill (1975) has discussed the ellipticity determination of Dicke and Goldenberg (1974), and has found that the claimed excess oblateness can be explained as due to a temperature and brightness difference between the equatorial and polar regions of the Sun. The dynamical ellipticity is consistent with a rigid, 27-day rotation period, while the oblateness may be as small as 8.6 (± 5.9) $\times 10^{-6}$ (not sensibly different from zero) on the measurements of Hill and Stebbins (1975; see also Hill 1975, Hughes 1975a). Even a value as large as the 4.5×10^{-5} of Dicke and Goldenberg does not support the scalar–tensor theory. This is because Schatten (1975) has used basic constraints, set by conservation of angular momentum and the torque exerted by the observed solar wind via magnetic fields frozen into the surface of the Sun, to deduce that the interior may be rotating quickly with a period of 1–4 days. This can produce an oblateness of up to 3.4×10^{-5} at the photosphere, and of a simple non-relativistic origin.

Thus several lines of study set an upper limit on the oblateness, and it is probably as small in practice as the values of order 10^{-6} quoted above. This means that the ellipticity actually present is ten times less than the value looked for by Dicke. Such a low solar ellipticity, to be consistent with the Brans–Dicke theory, implies a very large value for the coupling parameter, ω_{BD} (cf Hill and Stebbins 1975). A weak coupling also seems indicated by the time scale of pulsar energy loss by gravitational radiation, which is modified compared to the Einstein value for collapsed bodies ($2GM_*/c^2R \approx 1$) of reasonably large size. To be specific, a value of $\omega_{BD} \gtrsim 25$ can be inferred from planetary ranging experiments within the Solar System (Shapiro *et al* 1976) and this value is confirmed by laser tracking of the Moon, as discussed in §6.7. It is further confirmed by experiments on the deflection around the Sun of electromagnetic waves coming from distant sources such as 3C279. Modern versions of this classical test of general relativity can test the theory to 1 % accuracy (Riley 1975, Fomalont and Sramek 1976) and give $\omega_{BD} \gtrsim 23$. The data thus show support for Einstein's theory, as compared with that of Brans–Dicke, with agreement at the level of 0.99 (± 0.03) of the Einstein-predicted deflection. The similar test (often called the fourth classical test of general relativity) of measuring the time delay of an electromagnetic wave passing near the Sun has been applied using radar echoes from Mercury and Venus. The results are also in agreement with the simple Einstein-predicted value. This test has also been applied using the Mariner 6 and 7 spacecraft with an improved accuracy that gives agreement with general relativity to within 3% (see MacCallum 1976 for a short review of this and related experiments).

The quoted limits for ω_{BD} are considerably removed from the value $\omega_{BD} \simeq 6$ originally advocated by Dicke. This value is the one chosen to avoid an outright clash with the observed rate of advance of the perihelion of Mercury (the Brans–Dicke rate is $(4+3\omega)/(6+3\omega)$ of the Einstein rate) and yields an age of the Universe in agreement with data from astrophysics and radioactive decay measurements (Brans and Dicke 1961, Brans 1961). The quoted limits for ω_{BD} effectively mean that the perihelion advance of Mercury, the Sun's figure and the deflection and slowing of electromagnetic waves near the Sun are all still in agreement with Einstein's theory, and require at best only a very small scalar–tensor contribution.

Thus the long-continued controversy over the shape of the Sun as a test for the Brans–Dicke cosmology has reached a consensus in saying that no such new theory is required by solar physics. This clearly tends to cast serious doubt on the validity of the theory as a reasonable basis for the observable Universe. Nevertheless, perhaps one should consider other testable aspects of the Brans–Dicke theory before reaching a final opinion. In any case, many of the tests of this particular formalism are also valuable as tests of other variable-G theories.

Besides the perihelion advance of Mercury and the effect of the cosmology on the orbital motion of the planets about the Sun, and of the Moon about the Earth (Dicke and Peebles 1965), one of the older tests of the Brans–Dicke theory lay in the effects it predicted for radioactive decay schemes. The philosophy adopted by Dicke with regard to the Principle of Equivalence (Dicke 1957a, b, 1964a, b), leads to the expectation that the binding energy of nuclei should change with epoch. This is connected with the high accuracy of the Eötvös experiment in showing that the weight/mass ratio of a neutron is the same as for an electron + proton (Dicke 1957a). The predicted influence on decay times means that radiometric age determinations of, for instance, an old suite of rocks on the Earth should show a 10% difference as determined by uranium and strontium decay measurements. However, the data are probably not accurate enough to show a definite result, especially if the age of the Universe is much larger than 10^{10} yr. The last possibility, taking $G \propto t^{-1}$, would also compromise Teller's estimate of a high temperature on the Earth due to high solar luminosity ($\propto G^7$ or G^8), and make the theory generally very difficult to test due to the smallness of the effects to be looked for (Dicke 1957a, p357).

The effect of the theory on solar luminosity could be more accurately estimated if a close limit on $\dot{G}/G$ were available. A non-geophysical method of determining ancient values of G, that shows some promise of accuracy, has been proposed by Gilbert (1969), who used Newtonian cosmological theory in evaluating the effect of a change in G on galactic structure. Kanasewich and Savage (1969) have attempted to estimate what effect a similar change would have on nuclear structure. A comparison of α and β decay in radioactive nuclides does not preclude a variation in G if the age of the Universe is greater than 13×10^9 yr. Runcorn (1969a) advanced the suggestion that the 18·6 yr precession of the nodes of the lunar orbit could be monitored in ancient coral deposits. Since the latter are already known to mirror the number of days in the synodic month and the year (§5.4), the variation in these data would be dependent on G. Limits on $\dot{G}/G$ are indeed available from studies of the rotation of the Earth (§5.8), but the question of their consistency makes it problematical to estimate geological and solar effects from them.

Dicke (1962a) considers it probable that the Earth's surface temperature is proportional to (solar radiation flux)$^{1/4}$. The solar luminosity is taken to vary as G^8, so the Earth's surface temperature should vary as $G^{5/2}$ (the extra power of $\frac{1}{2}$ being due to the variation in the Earth's orbital distance). This dependence on $G^{5/2}$ gives the surface temperature of the Earth 500×10^6 years ago as 30°C. Dicke believes that the high temperature would have caused clouds to form, raising the planet's albedo and causing the temperature to stabilise (as in the same chain of processes as envisaged by Jordan (1962a) in considering the similar problem on Dirac's hypothesis). The high value of ancient temperature is not an embarrassment for the origin of

32

life, but the theory would be seriously compromised if the existence of Palaeozoic, or earlier, ice ages could be definitely established, as seems to be the case (§6.5). Dicke would appeal to radiation loss from the atmosphere caused by water vapour in this event, since the flux loss rate is proportional to $T^{3 \cdot 65}$, where T is the temperature.

There is considerable room for adjustment of the rate of undesirable processes in the Brans–Dicke cosmology. The scalar λ_{BD} considered at the beginning of this section changes in a way that is connected with the density of matter, ρ_U, the epoch of the Universe, t, and ω_{BD}, the dimensionless para-meter of the theory ($\omega_{BD} \simeq 6$ following Dicke), according to the equation

$$\dot{\lambda}_{BD} = -\frac{G_0 \dot{G}}{G^2} = \frac{8\pi G_0 \rho_U t}{(2\omega_{BD} + 3)} \qquad (2.16)$$

where G_0 is a constant (Dicke 1962a). For a flat space, with matter in the form of galaxies, the corresponding Hubble age is

$$T_H = \left(\frac{4 + 3\omega_{BD}}{2 + 2\omega_{BD}}\right) t \simeq \frac{3t}{2}. \qquad (2.17)$$

It may also be shown that

$$-\frac{\dot{G}}{G} = \frac{2}{(3\omega_{BD} + 4)t} \simeq \frac{1}{10t}. \qquad (2.18)$$

For $T_H = 12 \times 10^9$ yr, the age of the Universe is 8×10^9 yr, and $\dot{G}/G = 1$ part in 10^{11} yr^{-1}. The above equations show that the theory has considerable scope to adapt itself to any observa-tions, geophysical or otherwise, that can be made to test it.

The most stringent tests applicable are astronomical ones, such as those relating to oblateness and perihelion discussed above. Dicke (1962c) has examined other possible astrophysi-cal tests, noting that some globular clusters appear to be older than the galaxies to which they belong, and also older than the heavy elements they contain. These, and other dating prob-lems, Dicke would prefer to reconcile by using a cosmology in which $\dot{G} \neq 0$, the scalar–tensor theory being acceptable in this respect. Dirac's hypothesis would also be in agreement with the data if the reciprocal of Hubble's parameter was at least three times larger than the value of 13×10^9 yr current at the time Dicke was writing.

One gets the impression once again that there is no need for a cosmological solution to the questions posed by Dicke (1962c), since current astronomical data are just not accurate enough to allow any decision as to whether or not a problem of dating exists. The outstanding test of an astrophysical kind is, in fact, that mentioned in §2.3, due to Pochoda and Schwarzschild (1964), who calculated expected evolutionary paths for models of the Sun in which G is varying.

The approach adopted by Pochoda and Schwarzschild (1964) consisted in assuming $G = G_0(T/t)^n$, where T is the age of the Universe (taken in the range 8–15×10^9 yr). The exponent n may be varied from $n = 0$ (conventional theory, $G = $ constant), through $n = 0.2$ (a form of the Brans–Dicke cosmology, with the rate of variation as preferred by Dicke), to $n = 1$ (Dirac's hypothesis). The object of the exercise is to evolve a model with a certain T and X_H (hydrogen abundance) for a chosen n, until conditions are found such that the Sun will be in the state observed, and will not have turned off the main sequence and become a red object. Atomic time is employed, with t running typically from $T - 4.5 \times 10^9$ yr up to T. The Sun is assumed to obey the usual equilibrium equations with the proton–proton and carbon–nitrogen–oxygen cycles providing the solar energy, and the burning rate can be adjusted by varying the initial hydrogen abundance (X_H). Pochoda and Schwarzschild took the abundance of heavy elements as 0.04, in a model with a radiative core and convective envelope. The following results were obtained. For $n = 0$ ($G = $ constant), they found agreement with observations to require $X_H = 0.68$; for $n = 0.2$ ($T = 8 \times 10^9$ yr), $X_H = 0.72$ was needed, and for $n = 1$ ($T = 15 \times 10^9$ yr) $X_H = 0.80$ was necessary. In the last case, the value of T employed was found to be critical, since any smaller age caused the simple radiative core model to break down, indicating that the Sun had turned off the main sequence. The $n = 0.2$ case, if applicable, would mean that the Sun's luminosity, on average, has been different by a factor of two over its history compared with $n = 0$; while $n = 1$ would mean that it was five times brighter in the early stages of its evolution ($X_H = 0.80$). The sensitivity of the calculation is illustrated by noting

34

that for $n=1$, $T=15 \times 10^9$ yr and $X_H=0.76$ (instead of 0.80), turn-off occurred after only about 10^9 yr. Clearly, the Universe must be older than 15×10^9 yr if Dirac's hypothesis is valid, while the Brans–Dicke ($n=0.2$) cosmology is acceptable for all reasonable ages. The approach has been repeated by Gamow (1967b) with approximately the same results.

Sensitive tests of the rate of change of G such as the one described above are valuable, but it is likely that the viability of theories like that of Brans and Dicke will be decided by experiments conducted within the Solar System using artificial satellites and radar tracking. The rate of change of G usually quoted for the Brans–Dicke cosmology, $|\dot{G}/G| \simeq 10^{-11}$ yr^{-1}, is calculated for the pressure-free, flat-space case; the pressure-free, closed-space calculation gives a figure of about 3×10^{-11} yr^{-1}. Morganstern (1971a) has extended these estimates to pressure-filled flat space, and expects the Brans–Dicke scalar field to be noticeably important in collapse situations observed at the present epoch, and also in the early stages of Friedmann cosmologies, when the radiation density is greater than the matter density. Morganstern (1972) has also examined the pressure-filled, curved-space case, finding that positively curved spaces with pressure predict $|\dot{G}/G| \approx 10^{-10}$ yr^{-1}, a rate only just compatible with the 4×10^{-10} yr^{-1} limit set by Shapiro *et al.* The case of negatively curved, pressure-filled space is found to depend critically on the details of the cosmological model adopted. Morganstern (1971b, c, d) has examined the cosmological status of the Brans–Dicke theory and found some solutions in it which are the analogues of the Friedmann solutions in Einstein's theory.

While the variable-G aspect of the Brans–Dicke theory has been much discussed, less attention has been paid to finding exact cosmological solutions to the field equations. Some solutions have been found by Dehner and Obregón (1971, 1972), though most are of peculiar types in which $G \propto S^N(t)$, where S is the scale factor of the cosmology and N is a parameter of order unity. Solutions with $S(t)=0$ at $t=0$ do not admit negative curvature cases. One solution with $k=+1$ has a value of G that increases with time, and a coupling parameter $\omega_{BD}<0$, contrary to the usual form assumed by Dicke (ω_{BD} probably lies in the range $\omega_{BD}=3$ ($+5$, -1.7); Dehner and

Obregón 1971). The value of G changes as $S(t)$ changes, the main form studied by these workers requiring the rather high value of $\rho_U \simeq 7 \times 10^{-29}$ g cm^{-3} for the mean density of matter in the Universe. (See Lessner 1974 for a discussion and criticism of the work of Dehner and Obregón 1971, 1972.) More reasonable solutions to the Brans–Dicke field equations, perhaps of distinctive types, may well exist.

Dehner and Hönl (1969) have examined the cosmological and astrophysical consequences of the Jordan–Brans–Dicke theory with the object of determining whether $G \propto t^{-1}$ conflicts with astronomical observations. Their approach is via the parameter ω_{BD} of the Brans–Dicke (1961) theory, which can be varied from $\omega_{BD} = 0$ (when the radius of the Universe varies as $t^{1/3}$ and $G \propto t^{-1/2}$) to the limit $\omega_{BD} \to \infty$ (when Einstein's theory is recovered, with $G = $ constant). Since the energy output by nuclear reactions in stars depends strongly on $G (\propto G^7$ or $G^8)$, it is possible to obtain an expression for the luminosity of an object as a function of Hubble's parameter (taken as 75 km s^{-1} Mpc^{-1}) and redshift distance. In this sense, G varies in space since the redshift, z, is a function of distance. Distant objects should appear to be relatively brighter than those at intermediate distances, since their absolute luminosity is greater due to the G dependence, counteracting to some extent the effect of increased distance. By incorporating this property into the expected variation of brightness with distance, the bolometric magnitude of a normal galaxy composed of stars, each having a luminosity $\propto G^7 \times$ (mass)^5—after Teller (1948)— is found to be given by

$$\text{magnitude} = +18 \cdot 95 - 2 \cdot 5 \lg \left[\frac{(1+z)^7}{z^2} \frac{N_L(z)}{N_0} \right] \quad (2.19)$$

where $N_L(z)$ is the number of radiating stars, and N_0 is the total number of stars in the galaxy. A plot of this function, for galaxies comprising 10^{11} stars, follows the distribution of normal galaxies up to $z = 0 \cdot 2$, but then turns up and follows the distribution of radio galaxies and quasi-stellar objects (QSOs), which, with Seyfert and N-type galaxies, are inferred to form a continuous sequence. There is no significant difference between any of the possible cosmologies up to $z \simeq 0 \cdot 2$, but the curious fact emerges that the best fit to the data is obtained by adopting $\omega_{BD}=0$; the curves for $\omega_{BD}=\infty$ (Einstein) and $\omega_{BD}=6$ (the value preferred by Dicke for the Brans–Dicke cosmology) both proceed to areas of the plot where no objects are observed.

This is an unexpected result, as noted and discussed by Barnothy (1969), who also points out that the axes of the first two figures of Dehner and Hönl (1969) are wrongly labelled. The third figure of Dehner and Hönl is a plot of $(U-B)$ against redshift, which led Dehner and Hönl to inquire if the ultra-violet excess of QSOs as shown by their U and B colours can be explained on the Jordan–Brans–Dicke theory. The answer to this question is affirmative, within the large scatter present in all the data being used. Einstein's theory would ascribe the phenomenon to a real ultraviolet excess for galaxies beyond $z \simeq 0.2$, as opposed to its being a result only of the cosmology. In actual fact, the agreement is not good for any possible theory, and $(U-B)$ becomes insensitive to z for large red-shifts. The Jordan–Brans–Dicke theory with $\omega_{BD}=0$ has $(U-B)$ increasing as z increases, with the opposite trend expected for the Einstein case. The discrepancy is explained by Dehner and Hönl as being due to Rayleigh scattering of light by hydrogen atoms in intergalactic space (although the required density is very high by cosmological standards), and a spurious tendency which is introduced by assuming that the radiation from galaxies is black-body. Neither of these explanations is entirely satisfactory. Certainly, one is forced to conclude that the value of $\omega_{BD}=6$ originally advocated by Dicke for the Brans–Dicke cosmology is not in good agreement with the type of astrophysical data being discussed.

It appears, in fact, that there must be an error of judgment in the procedure of Dehner and Hönl outlined above. Barnothy (1969) was quick to show that it would result in the observable number of QSOs being about two for every galaxy observed at the same redshift distance. Since observations only reveal 10^{-8} QSOs per Mpc^3, and 5×10^{-2} galaxies per Mpc^3, this means that the results of Dehner and Hönl predict many more QSOs than are actually seen. Whether this discrepancy is due to the assumption of flat space by the authors (Dehner and Hönl 1969, p36), or else is due to numerical errors, is difficult to say. The method, since it is acceptable as a means of testing the

Jordan–Brans–Dicke theory and the magnitude–redshift relation in (flat) Brans–Dicke cosmology, has been more thoroughly investigated by Tinsley (1972). She takes into account (i) the effect of the deceleration parameter ($q_0 > 0.5$, since $q_0 = (2 + \omega_{BD})/(2 + 2\omega_{BD})$, and so it can differ from the usual flat Friedmann model having $q_0 = 0.5$); (ii) the effect of the main sequence turn-off point for old stellar populations, and (iii) the direct effect of a decreasing G. Luminosity proportional to G^7 is probably relevant for the light from most stars. The magnitude–redshift relation for the Brans–Dicke cosmology is found to be steeper than that for the Friedmann models, but acceptably so. A similar calculation for the Hoyle–Narlikar theory is discussed in §2.7.

The theory may therefore be acceptable from the viewpoint of astrophysical tests. It also survives limits set on the variation rate $\dot{G}/G$. These are of size $\simeq 2 \times 10^{-11} \text{ yr}^{-1}$ from data on the rotation of the Earth (Chapter 3), but less exact from direct determinations. The agreement of the theory with solar oblateness and precession results, on the contrary, is not good.

2.7. Hoyle and Narlikar

The views of Hoyle and Narlikar (1971) on the nature of mass are motivated to a large extent by a wish to explain certain anomalies in the redshifts of quasars and clusters of galaxies. In particular, quasars with redshifts differing by 20 000 km s^{-1} and more are observed to be connected by bridges, while others occur in clusters with conformable reddening. Such observations might be explained by the agency of a space-variable mass field. The redshift theory is then built on the foundation of conformally invariant combinations of the basic concepts of physics (mass, length, time, charge, etc), all reduced to one fundamental representation (taken to be length by Hoyle and Narlikar). Maxwell's equations are conformally invariant as they stand, but dynamics is not, because mass does not scale between different representations in an acceptable manner. To overcome this, mass is made dependent on the rest of the Universe by the agency of a mass field. Employing the wave equation of Penrose, a theory of gravitation is arrived at which tends to general relativity very closely for weak fields.

38

Mass becomes quadratically dependent on cosmic time by this approach, so that there is a finite redshift even though space is flat. If ϵ_{HN} is a dimensionless parameter (propagator) connected with a test particle of mass m in a mass field defined by M (where $m(X) = \epsilon_{HN} M(X)$, that is, position-dependent) it transpires that

$$\epsilon_{HN}^2 \, (n_p c^3 t^3)^{1/2} \approx 1 \qquad (2.20)$$

with an effective value of G that depends on the values of m and ϵ_{HN} (see §3.9; n_p is the particle density in space). For the basic equation (2.20) to hold at all times, as opposed to its being a chance coincidence peculiar to the present epoch, one possible solution is in terms of $G \propto t^{-1}$. A range of behaviour of $G = G(t)$ is actually possible, but $G \propto t^{-1}$ is chosen by the authors for the purposes of illustration.

While G is decreasing in the Universe, the average particle density, n_p, is tending to increase (but not as fast as in steady state theory), the total gravitational interaction being constant. The predicted (negative) value of $|\dot{G}/G|$ is 10^{-10} yr^{-1} (taking Hubble's parameter to be 5×10^{-11} yr^{-1}). This is consistent with the upper limit for the variation of $|\dot{G}/G| \leqslant 4 \times 10^{-10}$ yr^{-1}, set by Shapiro et al (1971), using data on the orbits of planets. Anomalous redshifts are explained by a locally fluctuating value of ϵ_{HN}, and the authors suggest that the Earth's radius should have increased over its history by 500 km or so, an amount probably smaller than that indicated by geological evidence (cf Chapter 6).

The latest form of the cosmology of Hoyle and Narlikar, and its application to geophysics, grew out of an earlier proposal for a conformal theory of gravitation (Hoyle and Narlikar 1964d, 1966d) and the need for the model Universe to be a perfect absorber in the Wheeler–Feynman sense. These two concepts are discussed at length in §3.9, since they have general applications. The mathematical framework, which is closely similar to that of Brans and Dicke, is given in more detail there. The cosmology is characterised by a properly integrated form of Mach's Principle, with the scalar curvature of the space–time being zero everywhere except at particles, where it takes the form of a δ function, which does not, however, give a singularity in the inertial mass at the particle. The theory had been previously criticised by McCrea (1965), Hawking (1965) and Pirani and Deser (1965). In particular, McCrea has objected that the sign of G in the conformal cosmology is just as arbitrary in choice as it is in general relativity, a disadvantage considering that all observations show gravitational interaction to be attractive. This criticism has been replied to by Hoyle and Narlikar (1966d) who point out that gravity always causes attraction in the new theory since inertial and gravitational mass always have the same sign. It is the conformal invariance that leads to this, since there can be no term like the Λg_{ij} appearing in some general relativity cosmologies. As previously mentioned, conformal invariance will be discussed at greater length in Chapter 3, but a short discussion is relevant here.

Conformally invariant theories are usually those in which the only allowed variables are the positions of interacting particles; this is a dictum basic to all so-called direct-particle theories, under which heading the cosmology of Hoyle and Narlikar must be placed. Particles in field theories such as electromagnetism are redundant, since they merely exist within the field as singularities. Fields are useful as a device to aid calculation in direct-particle theories, but it is the nature of the particles that is basic, and direct-particle formulations can be constructed that are equivalent to all known field theoretic interactions. The conformally invariant approach insists on invariance under changes of scale factors, analogous to changes of units, as well as the conventional invariance properties of general relativity. The scale factors may vary with position and time, but the frame of reference familiar to science is that in which all members of a given group of particles (protons, in fact) have the same mass, and in which general relativity is valid.

The conformally invariant gravitational theory has been applied to Friedmann models (Hoyle and Narlikar 1972a) and to the new model described at the beginning of this section (Hoyle and Narlikar 1971, 1972b). The cosmology of Hoyle and Narlikar was largely motivated by the wish to account for discrepant redshifts in astronomical objects such as QSOs, but there may be redshift anomalies much nearer the Galaxy. De Vaucouleurs showed some time ago that observations of redshifts in normal galaxies demonstrated a tendency to increase, according to galaxy type, in the order ellipticals→lenticulars→ spirals→Sc spirals (De Vaucouleurs 1972). Furthermore, Jaakkola (1972) has found systematic discrepancies in redshifts between different galaxy types, using an analysis of observations from selected rich clusters, groups and pairs of galaxies. These data possess an excess negative redshift residual for E, SO and Sa galaxies, and an excess positive residual for Sb and Sc galaxies, the discrepancies being of the order of 1 km s^{-1}. There are some grounds for criticising these data as involving selection effects, in the same way that the discrepant QSO redshifts in quasar pairs (Hazard *et al* 1973) have been criticised (Bahcall and Woltjer 1974) as being a chance projection effect. Lewis (1975) has concluded that the anomalous redshifts in many compact nuclei are probably due to familiar processes of astrophysical gas dynamics with no cosmological cause being required. On the other hand, Bottinelli and Gougenheim (1973) have confirmed Arp's claim that companion galaxies have, on average, redshifts that are 90 km s^{-1} greater than those of the bright parent galaxies, so that the theory retains some of its basic astrophysical motivation.

The astrophysical tests that can be applied to the Hoyle–Narlikar cosmology are similar to those used for the Brans–Dicke theory, although the G variation is stronger. Hoyle (1972a, b) considered the past variation of the luminosity of the Sun, extending the simple dependence on G^7 used by Dicke and others to considering the dependence at the bottom of the main sequence (luminosity $\propto G$), in the centre ($\propto G^7$), and at the top ($\propto G^4$). He concluded that the solar luminosity was three times the present one at an epoch $4 \cdot 5 \times 10^9$ years ago, but that the solar constant was five times its present value due to the Earth's orbiting nearer the Sun. The early evolution of stars in general must have been affected by a higher value of G and, in particular, the Chandrasekhar limiting mass should have been different, since it is dependent on $G^{-3/2}$ (see above). Hoyle supposed that the solar constant was three times its present value at an epoch 3×10^9 years ago, but avoided the extremely high temperatures this would create if the greenhouse effect were active by assuming that water vapour rose in great clouds over the Earth's surface, to condense and release their locked-up thermal energy as radiation into space. Taking the present mean temperature of the Earth as 280 K, Hoyle inferred that the temperature 3×10^9 years ago was 350 K (that is, 77°C). Living creatures would not find a temperature like this impossible to live with, since some bacteria can live at temperatures of 95°C, blue-green algae at 70°C, fungi at 60°C and animals at most temperatures lower than 50°C. Hoyle looked upon this progression as evidence that the evolution of life forms has been governed by the temperature, which is, in turn, governed ultimately by a changing value of G.

The rate of change of G is restricted by data on the deceleration of the Earth's spin (Chapter 5), and this is an important, if ill-understood, datum for variable-G theories. The situation as summarised by Hoyle (1972a, pp336–9) is as follows.

(a) Telescope observations give the tidal component of the secular change in the Moon's angular velocity ($n = d\phi/dt$, $t =$ atomic time in effect) as $(\dot{n}/n)_{\text{tidal}} \simeq -0.13 \times 10^{-9}$ yr^{-1}, the total deceleration being $(\dot{n}/n)_{\text{Moon}} = 2\dot{G}/G + (\dot{n}/n)_{\text{tidal}}$. What is determined is actually $(\dot{n}/n)_{\text{Moon}} - (\dot{n}/n)_{\text{planets}}$, involving measurements of the longitudes of planets, and one needs to know $(\dot{n}/n)_{\text{Moon}}$ before $\dot{G}/G$ can be calculated.

(b) Eclipse data, or at least some of them, agree amongst themselves in giving $(\dot{n}/n)_{\text{Moon}} \simeq -0.25 \times 10^{-9}$ yr^{-1}, in approximate agreement with star occultations observed against the Atomic Time service, which give $(\dot{n}/n)_{\text{Moon}} \simeq -0.29 \times 10^{-9}$ yr^{-1}.

(c) The estimates (a) and (b) together give $2\dot{G}/G \simeq -0.12 \times 10^{-9}$ yr^{-1}, so that $|\dot{G}/G| \simeq 6 \times 10^{-11}$ yr^{-1}.

(d) These results tend to be confirmed by the spin-down of the Earth's angular velocity (ω) which is $\dot{\omega}/\omega \simeq -0.26 \times 10^{-9}$ yr^{-1}, if (a) is correct, and is due to tidal effects.

(e) This value agrees with the ancient eclipse value of $\dot{\omega}/\omega \simeq -0.28 \times 10^{-9}$ yr^{-1}.

(f) If there were no term due to $\dot{G}$, or to secular global expansion, the $(\dot{n}/n)_{\text{Moon}}$ value of (b) would give $(\dot{n}/n)_{\text{tidal}} = (\dot{n}/n)_{\text{Moon}} = -0.25 \times 10^{-9}$ yr^{-1}. This would give $\dot{\omega}/\omega = -0.50 \times 10^{-9}$ yr^{-1}, by conservation of angular momentum, disagreeing with the eclipse value (e). The conclusion is that the rotation data (see Chapter 5 for a fuller discussion) are compatible with $\dot{G}/G \simeq$

42

-6×10^{-11} yr^{-1}, or with an equivalent process such as planetary expansion.

A test of the Hoyle–Narlikar theory which has given a fairly decisive result is that applied by Barnothy and Tinsley (1973). With regard to the Hoyle–Narlikar theory, they have analysed (i) the optical magnitudes of distant elliptical and spiral galaxies, which are effectively being viewed at an early stage in their evolution, when G may have been larger and the luminosity ($\propto G^7$, G^8) of the component stars greater, and (ii) the colours of remote ellipticals as predicted by the Hoyle–Narlikar theory. They find that neither (i) nor (ii) are in agreement with observation. The basic data used in this and similar studies concern how the luminosities and colours of stars change as G changes (Teller 1948, Roeder and Demarque 1966, Roeder 1967). Barnothy and Tinsley assume that changes in G affect only the stellar temperatures and luminosities, stellar luminosities being dependent on G by a factor somewhere in the range G^4–G^7. The predicted galaxies are much brighter and bluer than observations allow, showing that the theory is difficult to reconcile with fact, unless an ancient higher value of G was responsible for some fundamental difference in stellar structure compared with the present epoch. Taken in conjunction with the low limits set on G by studies of the Earth's rotation (see Chapter 5), the results of Barnothy and Tinsley tend to suggest that the Hoyle–Narlikar cosmology is incorrect.

A similar conclusion has been reached by Dearborn and Schramm (1974), who set limits on the variation of G by looking at clusters of galaxies. They employed a Plummer potential of the form $\phi_P = \phi_0[1 + (r/r_0)^2]^{-1/2}$ as a model for the gravitational field of a cluster. They then examined the expected effects of G variability, over the history of the cluster, using

$$
\left.
\begin{array}{l}
\dot{G}/G \propto t^{-1} \\[2mm]
G = G_i(t_i/t)^{\alpha_{BD}} \\[2mm]
G = G_i \exp\left[-(t - t_i)/\tau_c\right]
\end{array}
\right\}
\qquad (2.21)
$$

where α_{BD} is a parameter connected with the Brans–Dicke theory, τ_c is a characteristic time scale for the cluster and G_i and t_i refer to some initial epoch. The dynamics of clusters of galaxies enabled Dearborn and Schramm to obtain limits of $|\dot{G}/G| \leqslant 4 \times 10^{-11}$ yr^{-1}, if the cosmological deceleration parameter is $q_0 \ll 1$, and $|\dot{G}/G| \leqslant 6 \times 10^{-11}$ yr^{-1} for $q_0 = \frac{1}{2}$. If these limits are accepted at face value, they would rule out the simple Dirac dependency and also that advocated by Hoyle and Narlikar, but not that of Brans and Dicke. It is therefore important to note that Dirac (1975) has pointed out that the dynamical variability of G is a concept that can be discussed only when G (being dimensional) is combined with some other parameter to form a dimensionless number. Dirac is of the opinion that the study of cluster stability, as in the work of Dearborn and Schramm, gives no information whatsoever on a meaningful time dependency of G, since such a study involves purely mechanical laws. It is only in a comparison of the gravitational with the atomic time scales that such a dependency can be sensibly discussed. This procedure is correctly incorporated in the work of Van Flandern (1975) and Shapiro et al (1971, 1976). It does not seem to have been incorporated in the work of Lewis (1976) who, on the basis of numerical work with $G \propto t^{-1}$ dynamics, has claimed that the missing mass problem in clusters of galaxies can be resolved in variable-G cosmologies. Although not everyone is agreed on the meaning of variability in the dimensional constants (Chapter 4), the consensus of cosmological opinion is that Dirac's standpoint is valid.

The theory of Hoyle and Narlikar may therefore be cosmologically satisfactory, although if accepted, it would certainly alter prevailing concepts concerning certain peculiar astronomical objects, especially black holes. In another context, Hoyle (1969) has stated that he believes matter to emerge from areas of strong gravitational fields and not to collapse into singularities (this has connections with Jordan's hypothesis concerning the origin of stars). The constant of gravitation is necessarily positive, so weak gravitational fields are necessarily attractive in the Hoyle–Narlikar concept of mass fields. Although appealing, the theory runs into serious disagreement with observation in the realm of galactic astrophysics.

44

2.8. The Status of Variable-G and Related Theories

The Hoyle–Narlikar cosmology complements the Brans–Dicke theory (§2.6). Both are similarly constructed and represent the outcome of the Jordan theory (§2.4), which was in turn inspired by the Dirac hypothesis (§2.3), even though that theory has since been completed independently in a different manner. Almost all the theories considered depart from general relativity, the exception being Gilbert's attempt (§2.5) to allow for G variability within the formalism of Einstein's theory. The influence of Milne's ideas (§2.2) can be seen in all of the cosmologies, especially in the concepts of clock regraduation and the arbitrariness of units.

Although the theories of Milne, Dirac, Jordan, Gilbert, Brans and Dicke, and Hoyle and Narlikar are interesting in themselves, it must be clearly understood that a belief in the correctness of any one of them over general relativity represents a minority opinion. As with all departures from convention, it is likely that *some* of the component concepts that make up a new theory are the main contributions to an expanded understanding of the subject, not necessarily the whole theory. This is probably true of the cosmologies examined above. The ideas involving clock regraduation, coordinate and unit transformations, conformal invariance and the absorber theory of radiation are major concepts in themselves. There is a possibility that some or all of these ideas may ultimately have relevance for observational astrophysics. The same cannot be said of the complete theories, for the many reasons discussed above. While the Brans–Dicke and Hoyle–Narlikar theories are not in flagrant disagreement with observation, the former is compromised and the latter may easily become so. The Dirac complete theory, however, holds out the possibility of being correct in the usually understood meaning of the word.

In addition to the theories examined above, there are several others that might be mentioned without going into detail (Swann 1927, Chapman 1929; Holmberg 1956; Bailey 1960; Kapp 1960; Kalitzin 1967; Schwebel 1966; Aspden 1966, Aspden and Eagles 1972; Krat and Gerlovin 1974, 1975a, b; Fennelly 1974; Segal 1976, Nicoll and Segal 1974). Of the theories that have attracted notable attention, one might mention that of Prokhovnik (1970, 1971) which is based on ideas due partly to McCrea. The astrophysical motivation derives from arguments involving the missing mass problem and mass loss by gravitational radiation from the Galaxy (Sciama 1972). Further discussion of this and other theories are given in Wesson (1973), but none of them is compatible with general relativity. A modification of general relativity that may be conveniently mentioned here is the group theory of Malin (1974, 1975, 1976) and Mansfield and Malin (1976). Malin wishes to describe a cosmology by a de Sitter group and the particles it contains by a compatible Poincaré group at some epoch. The particles are labelled in spin and mass at that epoch with the eigenvalues of the Casimir operator in the associated space–time. This results in all particle masses being time-dependent on a Hubble time scale. The theory of Malin uses the field equations $R_{ij}=(-4\pi G/c^4)T_{ij}$, which were once considered by Einstein. He discarded them since $T^{ij}{}_{;j}\neq 0$, so there is no conservation of matter (the pressure is taken to be zero). It is difficult to see what justification there is for resurrecting this set of field equations, and in any case it is apparent that a theory of the Malin type could be arrived at much more simply merely by including a pressure term, perhaps due to Λ. As will be seen in Chapter 7, the inclusion of pressure causes bodies to lose mass in a way that can mimic that involved in Malin's theory.

Of theories that do not depart far from general relativity, the metric theories of gravity will be examined in §3.7. The other main class of theories are those quantum theories which cause slight modifications to, or require re-interpretations of Einstein's theory. These accounts usually begin from modern models for the elementary particles (see, for example, Wilkinson 1975, Blin-Stoyle 1975) and proceed to cosmological implications via work such as that of Domokos *et al* (1975; see also Davies 1976 and §7.6). They show that spontaneous symmetry breaking in gauge theories of the Salam–Ward–Weinberg type, as applied using Lagrangian theory to Robertson–Walker Universe models, can be traced to a cosmological origin (see §7.5). In such models, there was a phase transition in the big bang fireball, before which weak and electromagnetic forces had approximately the same strength and were long range. The weak interaction parameter has since decreased,

46

and its present rate of decrease is $\lesssim 10^{-10}$ yr^{-1}. With regard to broken gauge theories of the weak and electromagnetic interactions of the type put forward by Salam and Weinberg, it is becoming apparent that such formalisms can seriously affect cosmology via a term that corresponds to the cosmological constant Λ (see Weinberg 1974, Kirzhnits 1972). In particular, Linde (1974) has shown that spontaneous symmetry-breaking theories of elementary particles imply that the vacuum energy depends on the temperature of the cosmological medium. This is equivalent to having Λ depending on temperature, that is, on epoch, in such a model. It leads to the conclusion that Λ has changed from $\Lambda \approx 10^{-6}$ cm^{-2} in the early big bang, when $T \approx 10^{15}$–10^{16} K, to its present value of $\Lambda \lesssim 10^{-55}$ cm^{-2}. (Some workers quote values of Λ in dimensions length^{-2}, others in time^{-2}; the conversion parameter is merely c^2.) Such theories possess a finite value of Λ at early epochs and necessarily oblige one to take $\Lambda \neq 0$ at the present epoch. Furthermore, if any uncompensated charges exist (of a nature connected with the electromagnetic or weak interactions) in a space–time whose particle physics obeys the Weinberg model, then such a universe cannot be homogeneous, or isotropic, or possess a positive curvature and be closed. With regard to the last point, the belief is now gradually gaining acceptance that the actual Universe may indeed be open (Kirzhnits and Linde 1972). These results clearly imply that Λ can be an important term in cosmology, and that the Universe may possibly continue to expand forever. A last compelling argument for taking Λ to be finite is that only via Λ can the properties of the vacuum be included in a theory of the Einstein type, and this is probably true of all metric theories. These reasons for assuming $\Lambda \neq 0$, and for keeping a receptive mind to the idea that general relativity may need slight modification, will be discussed in Chapters 3 and 7.

In view of these comments, it is apparent that of all the theories, the Dirac theory is superior to that of Brans and Dicke and to the mathematically similar one of Hoyle and Narlikar. While all three theories remain viable at some level of approximation, it will become apparent in Chapter 3 that any new cosmology of this type implies profound changes in gravitational theory and should therefore be viewed with caution.

3. Gravitational Theory

3.1. Introduction

A theory of gravity can usually be stated succinctly in the form of an equation, or set of equations, from which the motion of a test particle near a given configuration of masses can be calculated. The concept is simple, but the large number of discarded theories of gravity is an indication of the practical difficulty of arriving at a theory that is in reasonable agreement with observation. Most theories that have appeared since Einstein's general relativity have possessed some aspect of their formulation that appeals to geophysics or astrophysics for validation. The criteria open to observational testing in such theories usually take the form of hypotheses concerning G, the Newtonian gravitational parameter.

In general relativity, it is well known that this parameter enters in terms of a quantity γ as follows. The 44 component of the Ricci tensor (R_{44}) and the energy-momentum tensor (T_{44}) are used in the weak-field approximation to Einstein's law of gravitation which, in the general form, is

$$R_{ij} - \frac{R}{2} g_{ij} + \Lambda g_{ij} = \gamma T_{ij}. \tag{3.1}$$

This results in an equation that is identical to Poisson's equation, $\nabla^2 \phi_{\mathrm{p}} = 4\pi G \rho$ (where ϕ_{p} is the gravitational potential, and ρ is the mass density of matter), provided the identification $\gamma = 8\pi G/c^4$ is made (see, for example, Lawden 1967, p154). This is one method of interpreting G. There is another that is of a somewhat different type, namely that occurring in the derivation of the Schwarzschild metric, which is frequently derived in texts by (a) defining a suitable action to use in the Euler–Lagrange equations for the variational problem; (b) working out the four Euler–Lagrange equations; (c) comparing

48

these equations with the standard equation of a geodesic, to give (*d*) the components of the Christoffel symbols. The latter can now be used (*e*) to calculate the components of the Ricci tensor, R_{ij}, which can then be (*f*) substituted back into the empty-space field equations as given by equation (3.1) with $T_{ij} \equiv 0$. This gives three new equations which can be solved for the unknowns that occurred at the start of the problem in stage (*a*). The last step involves the identification of a constant, *c*, as the velocity of light, and taking the conventional identification of *Gm* (*m* is the central mass) to be consistent with the Newtonian approximation to equation (3.1) above.

The procedure (*a*)–(*f*) results in the familiar Schwarzschild metric

$$ds^2 = c^2 \left(1 - \frac{2Gm}{rc^2} - \frac{\Lambda r^2}{3c^2}\right) dt^2 - r^2 (d\theta^2 + \sin^2 \theta \, d\phi^2)$$

$$- \frac{dr^2}{[1 - (2Gm/rc^2) - (\Lambda r^2/3c^2)]},$$

the conventional form of equation (2.12) where Λ is finite.

It is obvious, especially in the last instance, that what has been defined is the product of *G* and the mass *m* giving rise to the gravitational field. There is no possibility, in a given context, of *G* being variable, as, for instance, it would be if *G* depended on time. *G* is essentially a conversion factor that enters because mass is usually measured in other than gravitational units (see above). In fact, the product *Gm* is the only form in which *G* and *m* occur in practical orbit problems such as those of dynamical astronomy.

The complete Einstein theory shows that a logical distinction must be made between the *active* gravitational mass which, in a Newtonian sense, gives rise to a field that is felt by the *passive* gravitational mass. The concept of *inertial* mass, which is the mass characterising the resistance to acceleration and the energy equivalent of matter, completes the trio. The number *G* is the one free parameter that is a non-trivial comparison unit entering into a discussion of the three types of mass. Trautman (1964) has given operational definitions of the three types of mass in Newtonian theory, showing that inertial and passive gravitational mass can be proved to be equal as a theorem, provided Newton's laws hold, with the motion of a body taken to be independent of its constitution. It can further be proved that the ratio of active to passive (or inertial) mass is a universal constant (independent of the body), and this is conventionally labelled G. Its numerical value depends on the units employed ($G = 6.7 \times 10^{-8}$ cm^3 g^{-1} s^{-2} or 6.6×10^{-11} m^3 kg^{-1} s^{-2}), and it is often convenient to choose units in which $G \equiv 1$. In general relativity, all three types of mass are equal, but in certain other theories to be examined below, the inertial effects of the field and the matter interact, so that it is necessary to reparametrise G, or admit that it is variable.

3.2. The Origin of Inertia

The origin of inertia is usually assumed to be in some way connected with the inductive effect of distant matter, although the argument concerning the degree of incorporation of Mach's Principle into general relativity has been in progress since the formulation of Einstein's theory. Most of the complicated approaches to the problem reach only partial and unsatisfying conclusions. Sciama has investigated it from the simple viewpoint of an observer situated in Minkowski space with matter expanding outwards everywhere at the Hubble velocity ($v = r/t$). An observer to whom the redshifts of the receding galaxies appear isotropic is 'at rest' in this model universe, in which one imagines a particle to be moving (Sciama 1953), and on which one uses a theory analogous to that of electromagnetism. If a body of gravitational mass M is superposed on this uniform universe plus test particle, it generates a field in the rest frame of the test particle equal to

$$-\frac{M}{r^2} - \frac{\phi_p}{c^2}\frac{\partial v}{\partial t} \qquad (3.2)$$

where $\phi_p (= -M/r)$ is the potential of the body at the test particle. By postulate, the total field at the particle is taken to be zero to give equilibrium, so that

$$-\frac{M}{r^2} - \frac{\phi_p}{c^2}\frac{dv}{dt} = \frac{\Phi_p}{c^2}\frac{dv}{dt}. \qquad (3.3)$$

The scalar potential for all matter in the Universe (which is

50

of density ρ_U) is

$$\Phi_p \equiv - \int_V \frac{\rho_U}{r} \, dV = -2\pi \rho_U c^2 t^2, \qquad (3.4)$$

where the integral is carried out up to the light sphere over a volume V. From equation (3.3) one obtains

$$\frac{M}{r^2} = -\left(\frac{\Phi_p + \phi_p}{c^2}\right) \frac{dv}{dt}, \qquad (3.5)$$

which, as described in §3.4, has been interpreted by many workers as defining a space-dependent gravitational constant. Since $\phi_p \ll \Phi_p$, Sciama approximates equation (3.5) to give

$$G\Phi_p \simeq -c^2 \qquad (3.6)$$

with

$$\frac{1}{G} = -\left(\frac{\Phi_p + \phi_p}{c^2}\right). \qquad (3.7)$$

The total energy of a particle (inertial plus gravitational) at rest in the Universe is then zero. Combining equations (3.4) and (3.6) gives

$$2\pi G \rho_U t^2 \simeq 1. \qquad (3.8)$$

In this elementary approach to the origin of inertia, G is determined by the total gravitational potential at any point. With G as presently measured, and the present value of the time coordinate ($t = T$) since the big bang as $T = 10^{10}$ yr, the density is found to be $\rho_U \approx 10^{-29}$ g cm^{-3}. On the model of Sciama, 99% of local inertia is due to matter lying beyond 10^8 light years away, since there is a $1/r$ dependence involved. The fractional contribution due to the Galaxy is only 10^{-7}, so that $\phi_p \ll \Phi_p$ as used in equation (3.6) certainly holds, and is a justified approximation. Sciama (1953) notes that one can also obtain $G \rho_U t^2 \simeq 1$ from general relativity, but in the general relativistic case, G is not connected with the amount of matter in the Universe. Finally, the attractive nature of gravitation on Sciama's theory of inertia is assured, and is a consequence of the identically-equal-to-zero field at an arbitrary particle near a large mass.

Equation (3.6) of Sciama (1953) is not of unique derivation, since Bertotti (1962) obtained

$$\frac{G_0 M}{c^2 R} \simeq 1 \tag{3.9}$$

from a consideration of Mach's Principle. It was assumed that an inertial frame is one in which matter in the Universe is, in some average sense, at rest, so that

$$\frac{d^2}{dt^2}\left(\frac{\sum_i m_i \mathbf{r}_i}{\sum_i m_i}\right) = 0. \tag{3.10}$$

The sums extend to all masses in the Universe whose distances from a given particle are $\mathbf{r}_i$, with G_0 in equation (3.9) being the standard gravitational constant. To obtain the formulation of Mach's Principle contained in equation (3.10), it is sufficient to insist on equation (3.9) holding in combination with the statement that the total dynamical force on a body is zero. This is equivalent to equation (3.10), which states that a particle moves with uniform motion with respect to matter in the Universe. To see this, assume that $\mathbf{F}_N$ is the usual Newtonian force due to an acceleration $\mathbf{a}$, and $\mathbf{F}_t$ the tensor force due to a source of mass M, at a distance R from a test particle of mass m. Then the postulate of no net force requires that

$$\mathbf{F}_N + \mathbf{F}_t = 0 \tag{3.11}$$

where

$$\mathbf{F}_N = m\mathbf{a}, \tag{3.12}$$

$$\mathbf{F}_t = -G_0 \frac{Mm}{Rc^2}\mathbf{a}.$$

Equations (3.11) and (3.12) together give (3.10) if (3.9) holds. This type of analysis suggested to Bertotti that Coulomb's law should be modified to

$$m\mathbf{a} = \frac{e}{\sum_i (G_0 m_i/c^2 r_i)}\mathbf{E} \tag{3.13}$$

which is the same as saying that the charge-to-mass ratio of a particle is affected by large nearby masses ($\mathbf{E}$ is the electric field. The gravitational analogue of this phenomenon can be described by defining

$$\frac{1}{G} = \sum_i \frac{m_i}{c^2 r_i} \qquad (3.14)$$

so that the effective inertial mass of a body is $m_{\text{eff}} = mG_0/G$. This approach of Bertotti has a close affinity to that of Sciama, and both are similar to a related account by Barbour.

Employing M_U and R_U for the mass and radius of the Universe in equation (3.9), Barbour (1974, 1975) has sought to link Mach's Principle and the cosmic coincidence $G_0 M_U / R_U c^2 \simeq 1$ by developing a gravitational theory based on Lagrangians. He derives a relative-distance Machian theory (Barbour 1974) which accounts for the origin of inertia as the effect of incessant motion at the velocity of light of the particles within macroscopic bodies. Mach's Principle is then a coupling between such bodies, which can be expressed as the interaction of spinning matter rings (Barbour 1975). The theory has been reviewed by McCrea (1975c), and would seem to be compatible with general relativity and possibly with other theories of the Brans–Dicke type.

The three accounts just given of the relation $G_0 M / R c^2 \simeq 1$, as it might be connected with Mach's Principle, would all be of more firm standing if a definite criterion were available by which to evaluate the Machian properties of interacting masses. A reasonably clear statement and interpretation of Mach's Principle has been given by Raine (1975), who stated two Machian conditions: (a) a metric is Machian if, locally, it is a generalised inverse functional of the Riemann tensor, and (b) a space–time is Machian if the Weyl tensor is a linear functional of the matter source terms. These rather technical conditions physically express the belief that the metric should be completely specified by the matter present; the cosmological term involving Λ, for example, is to be classed as a non-Machian one, and should therefore be disregarded on this basis. Raine found that asymptotically flat space–times, vacuum solutions and spatially homogeneous cosmological models containing perfect fluids undergoing anisotropic expansion, or rotation, are all non-Machian. Conversely, Robertson–Walker models and pressure-free, perfect fluid models with spherical symmetry and Robertson–Walker (as opposed to

Heckmann–Shücking or Kasner) type singularities, are Machian. These classes leave open the possibility of some inhomogeneous cosmologies being Machian in the sense of Raine. Since even globally homogeneous and isotropic cosmologies have to admit that the local distribution of matter is not uniform, a discussion of Mach's Principle necessarily brings up the subject of possible anisotropies in the inertial effects of matter.

3.3. Inertial Anisotropy

Simple results on the origin of inertia lead to the expectation that asymmetries in the distribution of matter, and so of inertia, will manifest themselves in an apparent variability of parameters such as G. The dimensionless numbers of physics which are of order of magnitude unity have very tight limits set on them with regard to variation in space (Chapter 4). The same argument, based largely on the Equivalence Principle, does not apply to numbers like the gravitational coupling constant which are very small (Dicke 1959b, 1964a, p171). These numbers could well vary, and if local physics is to be influenced by the configuration of matter in the Universe, it is to be expected that such numbers might vary in space. To make useful predictions about this it is necessary to know how inertia is affected by anisotropies in the mass configuration of the Universe.

Speculations as to how asymmetric matter distributions might affect local inertia, via the kind of interpretation of Mach's Principle due to Sciama, have been made by Cocconi and Salpeter (1958). They considered $r^{-\nu}$ contributions by distant masses to local inertia. The mass m_G in our Galaxy would, for instance, contribute a fraction

$$\frac{\Delta m_I}{m_I} = \frac{m_G}{\langle r_G \rangle^\nu} \frac{(3-\nu)}{4\pi \rho_U R_U^{3-\nu}} \qquad (3.15)$$

to the inertial mass of local bodies (where $\langle r_G \rangle$ is the average distance from the Earth of the masses forming the Galaxy, R_U is the radius of the Universe, and ρ_U is the mean density of the Universe). If $\rho_U \approx 10^{-29}$ g cm^{-3}, $R_U \simeq cH^{-1}$ and $\nu = 1$ (as usually assumed), then $\Delta m_I/m_I$ is of the order of one part in 10^6 due to the matter near the centre of the Galaxy. For $\nu = 0.25$, it is one part in 10^{10}. Anisotropies in local inertial mass would shift energy levels of orientated atoms and broaden lines in the spectra of certain nuclei. Tests of these effects have been made by Cocconi and Salpeter (1960) and Sherwin *et al* (1960). The most sensitive test was that of Hughes *et al* (1960), which placed an upper limit of 10^{-20} on $\Delta m_I / m_I$. This limit depends on certain assumptions regarding the response of ^{7}Li to nuclear magnetic resonance, and presents an interesting problem because the noted limit is less than the $\nu \to 0$ anisotropy expected for the Galaxy. This, and similar tests which give $\Delta m_I / m_I \lesssim 10^{-10}$, certainly indicate that the anisotropy in inertial mass at the Earth due to the Galaxy is very small.

Effects of $\Delta m_I / m_I$ can be expressed in terms of a variable G as discussed in §3.4, and such changes would show up in slight variations of other parameters, such as the fine-structure constant (Bertotti *et al* 1962, p12). If the fine-structure constant α ($\equiv e^2 / \hbar c$ in CGS units; $e^2 / 4\pi\epsilon_0 \hbar c$ in MKS) depends logarithmically on G as suggested by Landau (1955, see §4.8), it can be calculated that a secular change in G of 10^{-10} yr^{-1} would give a change in α of 10^{-12} yr^{-1} (Landau 1955). A yearly modulation of G of one part in 10^8, as suggested by several workers, would lead to a periodic change in α of one part in 10^{10} (see below). Since very strict limits on the time and space variability of α exist, it is fortunate for the Brans–Dicke cosmology that Dicke (1960a, 1961b, 1964a) has been able to devise an argument based on Mach's Principle to the effect that no mass anisotropy influences should be observable. The experiment of Hughes *et al* (1960), which was also independently performed by Drever (Dicke 1964a, p178), in Dicke's view, sets precise limits primarily on boson fields, these being such as to permit a choice of coordinate system for which physical laws appear locally isotropic in space. While the inertial effects of matter may vary as $1/r$, the role of Mach's Principle in the expected relationship turns out to be somewhat empty. This is not to say that the role of Mach's Principle as a basic postulate and guiding assumption of the Brans–Dicke theory is without justification.

The theory is, in fact, largely based on Mach's Principle (Dicke 1958, 1959b, 1962b, c). Imagine, for instance, that there is a spherical shell of matter with an observer inside it. On the basis of general relativity (or Newtonian theory), all the Riemannian invariants inside the shell are zero and space–time is flat (Dicke 1962b, p37). To Dicke, this is unacceptable because, in the spirit of Mach's Principle, it is to be expected that the matter in the shell will *somehow* manifest its presence to the observer inside the cavity. To incorporate Mach's Principle into physics, one somehow has to get around the fact that the Strong Principle of Equivalence and the Eötvös experiment as usually understood demand the constancy of all physical constants, including G, in time and space (Dicke (1959b) has discussed experimental aspects of the Eötvös experiment). While there is evidence in favour of this principle holding for α, there is yet some doubt, as stated elsewhere, whether it holds for the two weak interactions, gravity and β decay.

In the case of gravity, in fact, it can be demonstrated from the spherical-shell model that some degree of variability is to be expected. Within the spherical shell of matter, it is possible for a physicist to modify the value of G he measures by moving around inside the shell. If there were no other matter in the Universe, the variation could be quite large. The easiest way to see this is to consider the acceleration of the Earth toward the Sun, which by dimensional analysis (Dicke 1962b, p33) might be expected to be

$$a \simeq \frac{M_\odot}{R^2} \frac{1}{\rho_U T^2},$$ (3.16)

where ρ_U and T are the present density and age of the Universe, and $M_\odot$ and R are the mass and distance of the Sun. The presence of parameters relating to the Universe in equation (3.16) is an explicit recognition of Mach's Principle. From equation (3.16) one might identify

$$G \propto \frac{1}{\rho_U T^2},$$ (3.17)

which could be expressed in a form like equation (3.14) as

$$\frac{1}{G} \propto \frac{M_U}{R_U c^2}$$ (3.18)

where M_U is the mass of the Universe out to the horizon and

R_U is the radius out to the light sphere (defined by $R_U \equiv cT$). The value of G is now seen to be dependent on the mass of the Universe. However, Mach's Principle really leads us to expect that it should be the inertial reaction of matter that depends upon the mass distribution of the Universe, and so the inertial mass (m_I) of a particle should really change with position, not G. This reappraisal can be easily accomplished by changing the unit of mass used to express ρ_U. The change in units required is an important aspect of the theory; in changing the unit of length, for instance, one could allow $\hbar/m_e c$ or $(G\hbar/c^3)^{1/2}$ to vary with position, but not both. In the former case, particles do not move on geodesics, while in the latter, $(G\hbar/c^3)^{1/2}$ is not an invariant. The most convenient choice is that in which particles move on geodesics with fixed masses, but feel a variable G (Dicke 1962b, p34), and it is in this form that the Brans–Dicke theory is best known.

The space variability of G is closely connected with the use of unit transformations, and it is desirable to consider them briefly before proceeding to enumerate results on the space variability of G. (A discussion of conformal transformations, which are basic to the Hoyle–Narlikar theory, is given later in this chapter.) The laws of physics must be invariant under a transformation of units (Walker 1945) and the question of admissible units and coordinate transformations is not a trivial one. McVittie (1945) has examined the regraduation of clocks in spherically symmetric space–times of general relativity, showing that a regraduation is equivalent to a coordinate transformation within space–time and not to a change from one Riemannian space–time to another. Narai and Ueno (1960) have shown how the regraduation of scales and clocks can convert, for example, Schwarzschild space–time to Minkowski space–time by the use of a coordinate transformation with re-identification in the new metric of quantities such as the velocity of light.

Coordinate systems should be held fixed under a units transformation, whereas under a general coordinate transformation the system of physical units is fixed, but the coordinates are varied. Under a general transformation of units, therefore, the labelling of space–time coincidences between pairs of particles (with given coordinates) is invariant. On the other hand, the scalar curvature and other purely geometrical scalars, while invariant under coordinate transformations, are generally not invariant under a transformation of units. The motive for allowing various types of transformation can be discerned in the impossibility of, for instance, comparing the size of an atom on Sirius with the size of an atom on the Earth. The invariance of the laws of physics under coordinate-dependent transformations of units has been discussed by Dicke (1962d) with special relevance to Mach's Principle. This results in changing the variable-G form of the theory to the alternative form in which G, h and c are identically constant, but the rest masses of particles vary with position as a function of the scalar field, ϕ_{BD}, although the mass ratios of particles are still invariant.

The method of carrying out the units transformation is relatively straightforward. If one scales the unit of time (say) by a factor ξ_s, the metric tensor, g_{ij}, and interval

$$\mathrm{d}s = (g_{ij}\,\mathrm{d}x^i\,\mathrm{d}x^j)^{1/2} \tag{3.19}$$

scale by powers of ξ_s depending on their dimensionality. The masses of particles scale similarly. Putting the new quantities into the original theory in place of the old ones results in a new formalism in which Einstein's field equations are satisfied. Dicke also transforms Riemannian space to a flat space by redefining measures of time and length along three mutually perpendicular axes. This involves the velocity of light apparently being a function of position and direction, so that space is now anisotropic (Dicke 1962d). Considerations of this type bring in the possibility of a space variability of G.

3.4. The Space Variability of G

Anisotropies in the inertial effects of non-uniformly distributed matter can be expressed in some theories—via suitable transformations, as noted above—in terms of gravitational asymmetries. With the variable-G formulation, the variation in space of the m_{I} of a particle (Dicke 1962b, pp26–31) appears as space variation of G. The energy of a particle is still equal to $m_{\mathrm{I}}c^2$, although this would not hold if there were two tensor fields in gravity (unlike the one tensor and one scalar in the

Brans–Dicke theory). The experiments of Hughes *et al* (1960) and others to measure the anisotropy of mass show that it would be very difficult to incorporate more than one tensor field into gravitational theory (Dicke 1962b, pp30–1). To do so would make particle masses and α space-variable, which are undesirable consequences. In the Brans–Dicke theory, the value of G decreases because the matter of the Universe is receding (Dicke 1962b, pp36–43). Whether the redshift of the galaxies is due to recession or due, as some have claimed, to a time dependence of atomic frequencies owing to the emitting atoms becoming smaller with time, is to this extent irrelevant. Observationally, the two hypotheses are equivalent, the distinction being bound up with the question of choice of units.

G may also vary as the Earth pursues its orbit around the Sun (Dicke 1962b, pp22, 43), the variation expected being an annual one of order 10^{-10}–10^{-11}. This space variability arises because in scalar field theories of gravity the ratio of gravitational to inertial mass for a planet differs from unity by an amount $\approx$ (gravitational binding energy/rest mass energy). In the Brans–Dicke theory, for example, Jupiter's acceleration in a gravitational field would be less (by a term of about two parts in 10^8) than the corresponding acceleration of a small test body. Such differences from general relativity can, in principle, be used to test scalar field theories by observations in the Solar System (see Zeldovitch and Novikov 1971, p79). The possible space variability of G in this manner might be expected to give an annual change of 10^{-13} for $\delta\alpha/\alpha$, but such an annual variation is too small to be detected at present (Dicke 1962b, p22). The change in G for bodies that are strongly self-bound gravitationally and move in space (such as Jupiter; see Dicke 1962b, pp18–9) might be detected in the future if the orbital characteristics of the bodies of the Solar System can be determined accurately enough.

It will be recalled (equation (3.7)) that Sciama's analysis of the origin of inertia gave rise to the formulation of a space-variable G (Sciama 1953). This has been interpreted (see, for example, Bertotti 1964) in the form

$$G = G_0 \left(1 + \frac{a'\phi_p'}{c^2}\right) \qquad (3.20)$$

where a' is a constant and ϕ_p' is the gravitational potential in the region of the body or bodies being considered. The relation in equation (3.20) has been commented on, in passing, by several workers (Einstein 1950, Jordan 1955, 1959, Brans and Dicke 1961, Dicke 1962c, d), but the most notable attempt to interpret it practically has been made by Finzi (1962), who investigated the variation of the gravitational self-energy of a body as G varies. This gives rise to a force on a white dwarf which may imply significant corrections to the orbit it pursues in the Galaxy; the path is expected to be slightly different from that of a normal star. In particular, white dwarfs in very weakly bound clusters will escape, while those in more strongly bound clusters will occupy anomalous positions.

While the proposed tests of equation (3.20) due to Finzi (1962) still await application, Steiner (1967) has made some highly speculative comments concerning orogenies, palaeomagnetism and other geophysical effects. He believes that the latter occur with a 350×10^6 yr period, which he justifies by appealing to the variation in G as the Earth pursues its elliptical orbit around the Sun and the Sun its orbit about the Galactic centre. A similar proposal has been made by Machado (1967, 1975), but the grounds for his claim are incorrect. The interest in periodic changes in G that lead to periodic changes in the Earth's radius according to the proportionality $|\delta R_E/R_E| \propto |\delta G/G|$ has caused some comment on ways in which such changes might be detected geophysically (Stewart 1970, Dubourdieu 1973, 1975, Smith 1975a). There is evidence for a 200×10^6 yr period, based on changes in sea level, but such a cycle could easily have a non-cosmological origin, such as that suggested by Ilič (1974) involving geosynclinal activity.

I do not consider that any of the above arguments, or those involving the periodic effects of a changing G on the Earth, have empirical support. Theoretically detailed bases are lacking for such proposals, with the exception of a good account by Rochester and Smylie (1974) of the changes in the trace of the Earth's inertia tensor (as the Earth pursues its orbit) predicted by certain metric theories of gravity. These workers have combined the possibility of changes in

$$\sum_{i=1}^{3} I_{ii} = \mathrm{Tr}(I)$$

60

with the results of Lambeck and Cazenave (1973) on the contribution of atmospheric winds to seasonal variations in the
length of the day, to obtain the result that changes in G produce
changes in the inertia as given by

$$\frac{\Delta\mathrm{Tr}(I)}{3I_{33}} \simeq -\frac{\Delta G}{10G}.$$

From geophysical limits on changes in the inertia of the Earth
and the coefficient J_2 of the geoid, they obtain the limit

$$\frac{|\Delta G|}{G} \lesssim 2 \times 10^{-9}$$

for allowable changes in G (in the sense of a spatial gradient
of G across the orbit of the Earth as predicted by preferred-
frame theories of gravitation). This geophysical limit is better
by a factor of three than previous ones set by other means.

Since there are reasons for believing that G might be space-
variable as discussed above, it may be useful at this point to
quote some rough limits on the possible dependency of this
parameter on the sizes of astrophysical systems. In astrophysics, there is a puzzlingly large number of types of object in
which the laws of dynamics, as used locally with $G=G_1$,
do not seem to hold accurately. For this reason, it is of speculative interest to carry out a short survey of such systems and
the possible reasons for their anomalous behaviour. In so far
as these anomalies could be made to disappear by preserving
the equations of dynamics in their familiar forms, but altering
G, one can note the following.

(a) The well known problem of hidden, or missing, mass in
clusters of galaxies (if one rejects the much-criticised quick-
disintegration, black-hole binding mass, or massive-halo
hypotheses), can be accounted for if $G(\text{clusters}) \approx (10\text{--}100)G_1$.
This is based on the use of the virial theorem with an observed
cluster mass, M_o, a mean dispersion velocity, σ_v, and a mean
cluster radius, R_c, so that $M_o \approx \sigma_v^2 R_c / G$ is satisfied.
(b) In the Galaxy, the sizes of HI clouds, as analysed to be stable
against perturbations using the virial theorem, should have
sizes $\approx GM/3R'T$, where M and T are the mass and temperature
of a cloud (which are related via the observed density) and

61

R' is the gas constant. The predicted size is often an order of magnitude different from the observed size ≈ 5 pc, which might suggest that G (clouds) $\simeq 10\, G_1$.

(c) The Local Group of galaxies is dynamically dominated by M31 and our own Galaxy, these two objects having an anomalous velocity of approach of $\simeq 100$ km s^{-1} (Oort 1970) that can be brought into conformity with the conventional two-body prediction either by a hidden-mass hypothesis, or by having G(Local Group) $\simeq 7\, G_1$.

(d) There are numerous aspects of the theory of galaxy and star formation by gravitational instability that would be elucidated if G for those systems were larger than G_1.

(e) Associations of O-stars, although spherical in shape, often have densities too low to prevent disruption, this being one example of the general observation that astrophysical systems often appear to have positive total energy (Ambartsumian 1958). The O-star associations are particularly notable in this respect, but could be stabilised if G(O-stars) $> G_1$.

(f) Variable stars such as Cepheids can be observed in galaxies at some distance away from the local region of space. It is possible, in theory, to bring into alignment a large range of variable-star phenomena by adjusting the value of G, since the (usually) virial-derived frequency of pulsation and luminosity variation, ω_L, depends directly on G. This can be seen by imagining a sound wave to propagate across a star of radius R_*, at speed v_*, the pressure and density being p_*, ρ_*, and the gravitational energy $|W_*|$, connected with the internal energy, U_*, by $3(\gamma_* - 1)U_* \simeq |W_*|$. Stability against perturbations is assured if the ratio of specific heats is $\gamma_* > 4/3$. By basic physics

$$\omega_L^2 = \frac{v_*^2}{R_*^2} \simeq \frac{p_*}{\rho_* R_*^2} \simeq \frac{U_*}{\rho_* R_*^5} \simeq \frac{|W_*|}{\rho_* R_*^5} \simeq \frac{G(\rho_* R_*^3)^2}{\rho_* R_*^6} \simeq G\rho_*.$$

$$(3.21)$$

While the large and disordered group of phenomena usually referred to as flare-originating (Lovell 1971) can be made more consistent by allowing spatial variations in G, it is important to note that there is evidence (Sandage 1972) that *some* Cepheids in different galaxies have nearly the same properties as those in our own Galaxy. This places a constraint on any theories that lead to considerable G variability in space.

(g) Active galactic nuclei and quasi-stellar objects provide grounds for speculation on the possible variability of G, because their large luminosities—if thermonuclear in origin—depend on G^7 or G^8, and so can be explained by values of $G > G_1$. In addition, the acceleration and confinement mechanics of large populations of energetic particles necessary to give synchrotron radiation can be modified by changing G arbitrarily. The most feasible effect of a space-variable G in QSOs, however, lies in their multiple redshifts, which Karlsson (1971) analysed in an attempt to clear up the controversy about their non-random distribution. He decided that the redshifts were non-random and invoked a space variation (quantisation) of the electron mass m_e. Subsequently, the non-randomness has been largely shown to be a selection effect by Weymann *et al* (1978) and others, but the multiplicity of redshifts within one object is a phenomenon which might be explained by some hypothesis involving a jigsaw-like matching of space–time domains (Ne'eman and Tauber 1967). Such joining of domains has been considered by Synge (1966, p344). The joining of the domains in the general case involves certain continuity properties of the metric tensor, g_{ij}, and its derivatives. Synge (1966 pp39–41) has given a good account of the joining conditions (see also Pirani 1962, p95, Harrison *et al* 1965, pp71, 127, 133–4). The problem is reminiscent of a Universe built by Lindquist and Wheeler (1957; see also Wheeler 1962, 1964a, pp369–87, 1964b, pp228–32) out of a large number of juxtaposed and appropriately joined Schwarzschild metrics. The relevance of constructions of this kind to peculiar objects is not completely understood, but the method will be met with again in Chapter 7.

The preceding astrophysical instances (a)–(g), for which a space variability of G could be proposed, do not, in my opinion, show any positive aspect. On the contrary, (f) provides an (admittedly uncertain) indication that G does not vary by any very large amount with astronomical location. On the other hand, it cannot be denied that G, among all the physical constants, is outstanding in being very badly determined by experiment compared with the other parameters such as c, h, etc. Cohen and Du Mond (1965) have reviewed methods of determining the values of the constants of physics by experiment, but even the most refined method of determining G (Rose *et al* 1969) still leaves it strangely inexact. This peculiarity is well known in astronomy as being the source of annoying restrictions on the calculations of dynamical parameters and orbits. A method of Beams (1971) holds out some prospect of improving conventional accuracies of the order of five parts in 10^3 and obtaining G exact to one part in 10^4. Beams' method might also be used to measure the time rate of change of G in the laboratory, while Braginskii and Ginzburg (1974) have suggested methods involving pendulums and gravimeters that could detect a rate of $|\dot{G}/G| \approx 10^{-11}$ yr^{-1}. Such methods as these, however, still await instrumental developments.

3.5. Vacuum Elasticity and G

The influence of matter on the nature of space–time can be approached in a more direct way than that represented by Mach's Principle, by following the approach adopted by Sakharov and others (see Zeldovitch and Novikov 1971, pp71–4). This method, unlike those of Sciama, Bertotti and Dicke, is based on general relativity. Sakharov has identified the coefficient of the second term in the Einstein field equations as being associated with a kind of elasticity of the vacuum. Deriving the equations from an action, S_A, gives the form

$$\frac{\delta S_A}{\delta g^{ij}} = \frac{T_{ij}}{2c} - \frac{c^3}{16\pi G}\left(R_{ij} - \frac{Rg_{ij}}{2}\right) = 0 \qquad (3.22)$$

where the symbols have the usual meanings (g_{ij} is the metric tensor, R_{ij} is the Ricci tensor, R is the curvature invariant, and T_{ij} is the energy-momentum tensor). Equation (3.22) has a form similar to ones in continuum mechanics with an elastic restoring term involving G. The constant of elasticity, $c^3/16\pi G$ is very large, so that the curvature of space produced by an elementary particle, for example, of mass m, is very small, the particle being spread over its Compton wavelength h/mc. The curvature produced is very small measured in units of length $h/m_\mathrm{p}c$, where m_p is the proton mass. The fundamental atomic unit of mass, for comparison, is large: $(\hbar c/G)^{1/2} = 2 \times 10^{-5}$ g (Zeldovitch and Novikov 1971, p66). The object of Sakharov's work is to obtain the second term in Einstein's equations from the curved-space equations of action as contained in the first term. The T_{ij} also specify the quantum aspects of matter which are required to connect up with G. This is done by introducing a momentum cut-off (p_0) into quantum theory, such that the correction due to the action, S_A, gives the elasticity correction as

$$\frac{c^3}{16\pi G} \int R \, dV = \frac{k_A p_0^2}{\hbar} \int R \, dV \qquad (3.23)$$

where k_A is a dimensionless constant of order unity and dV is a volume element. The choice of p_0 as specified by the fundamental mass, 2×10^{-5} g, gives

$$\frac{m^2 c^2}{\hbar} = \frac{c^3}{G} \qquad (3.24)$$

which is to be read from right to left, and intrinsically defines G in terms of p_0, or the equivalent mass, m. The justification for the choice of p_0 is that it gives approximately the correct value of G, the correct elementary particle charge, and the correct weak-interaction constant.

Sakharov's assumptions seem to be acceptable, except in taking the cosmological constant Λ in Einstein's equations as $\Lambda \equiv 0$. There are reasons for believing that Λ is in some way connected with the physical properties of the vacuum (Anderson and Finkelstein 1971, Zeldovitch and Novikov 1971, p29), although it is undoubtedly not easy to couple with the other physical constants (Zeldovitch and Novikov 1971, p73). This does not seem to me to be a reason for putting $\Lambda \equiv 0$, but rather an excuse. It will be seen in Chapter 7 that the consistency of interpretation of a certain class of relativistic cosmological models connected with geophysics and astrophysics requires that Λ be finite (cf §2.8). The acceptance of Λ as the manifestation of physical properties of the vacuum, while not perhaps a hypothesis that is easily acceptable, might also help to elucidate the curious symmetry, first noted by Epstein (1973), between a massive sphere situated in a vacuum and a vacuous sphere surrounded by matter. Epstein realised that the force of attraction between two vacuous bubbles that displace masses m_1 and m_2 of matter is given by Newton's law, force $= Gm_1m_2/r^2$, when the bubble centres are at a distance r apart in a material medium. This matter–vacuum symmetry has been discussed in depth. Gross (1974) pointed out that the equivalence of the corresponding inertial and gravitational masses would probably not hold, for the inertial mass would depend on the shape and direction of motion of the body, in the case where matter and vacuum are interchanged. McCrea (1975a) has discussed the symmetry from a general relativistic point of view, with Λ taken as determining the zero levels of density and stress in Einstein's theory. A further argument in favour of taking Λ to be finite is sometimes made on the ground that any general relativistic method for synthesising the fundamental constants into a deductive whole ought to incorporate a non-zero cosmological constant.

3.6. Anisotropy and the Eötvös Experiment

The laws of physics need to have the constants, such as G, e, h, m_e, m_p and so on, substituted numerically before experimental results can be made to balance. Although the absolute values of the dimensional constants depend on whether one uses CGS, FPS or MKS units, the dimensionless numbers, for example, $\hbar c/Gm_p^2 \approx 10^{40}$, are true constants in conventional theories. They are independent of space and time, as assumed in the promulgation of the Strong Principle of Equivalence of general relativity. On the other hand, the Weak Principle of Equivalence has been considered closely in connection with variability of the parameters of physics, and can be used to set limits on such variabilities. The invariance of the fundamental constants is required by the Strong Principle of Equivalence, but the latter is so far only directly upheld by an elaboration of an argument of Wapstra and Nijgh concerning the gravitational redshift of light (see Dicke 1964a, p169). Experimentally, the Weak Principle of Equivalence, which is essentially a formulation of Galileo's experiments with falling masses of different substances, has more experimental support in the form of the Eötvös experiment. This relates gravitational acceleration to the mass of a body, and can be used to formulate indirect limits on the variation of some of the dimensionless constants. For instance, if the fine-structure constant, α, varies near a large mass such as the Sun $(M_\odot)$, some cosmological arguments might lead one to expect that the local value of α would depend on the quantity $GM_\odot/rc^2$ (Dicke 1964a, p171), where r is the radial distance. The size of such a variation in α, however, would lead to an anomalous acceleration much greater than the limit set by the null result of the Eötvös experiment, so the possibility of such a variation can be effectively ruled out.

It seems likely that the Eötvös experiment, being a null experiment, will provide even more stringent limits than those already available on possible variations in the parameters of physics. Accuracies of greater than one part in 10^{12} for the equality of gravitational and inertial masses are expected to be attained (Haugan and Will 1976), and preliminary results suggest that the Principle of Equivalence will prove to be upheld to this level of accuracy. The limits of order one part in 10^{11} already established for the Eötvös experiment also enable the Principle of Equivalence to be tested for the energy of the weak interaction, this being important for various metric theories of gravity. The weak interaction does obey the Equivalence Principle to one part in 100 (Haugan and Will 1976), and Einstein's theory clearly continues to gain support from the null results of the Eötvös and related experiments.

It has been claimed by Schiff (1959, 1960) that if the Eötvös experiment were to give an *exactly* null result, then none of the dimensionless coupling constants could vary with position, on an argument relating binding energy and work done in different places in a gravitational field (Bertotti *et al* 1962, p9). This is important for the space variation of G, as G is essentially a comparison constant between passive gravitational and inertial mass. While the concepts of passive and active gravitational masses can be combined in general relativity and most other well founded metric theories, in some other contexts they cannot. However, the active and passive gravitational masses are known by experiment to be equal to an accuracy of 5×10^{-5} in bromine and fluorine, as can be calculated from the Kreuzer (1968) experiment as discussed by Will (1976). For the rest of this chapter, attention will therefore be concentrated on the

Eötvös experiment and the question of the equality of passive gravitational mass and inertial mass.

Denoting these two types of mass by m_P and m_I, Schiff (1959) was able to reach some interesting conclusions regarding the nature of matter and energy. In particular, he obtained results on antimatter that may be summarised as follows: $m_P(e^-) \simeq m_I(e^-)$ to one part in 10^4; $m_P(e^-) \simeq m_P(e^+)$ to one part in 100, both particles having the same mass sign, so that matter and antimatter are inferred to behave similarly in a gravitational field. The symbol m_P should not be confused with that for the proton mass (note that Israel and Khan (1964) have discussed negative masses in general relativity). The second result also holds for nucleons and antinucleons, the proof depending on the consistency of Hamiltonian quantum theory. Schiff also noted that violation of the Principle of Equivalence would imply violation of the conservation of energy. There is no reason to doubt the latter principle, and any evidence of its breakdown would be in contradiction to general relativity, where energy conservation is established as a tensor identity.

Similarly, there is no evidence for the non-conservation of the P, C and T symmetries in gravity (Hari-Dass 1976). The latter are all broken in elementary particle theory, although the product PCT is conserved. Dirac's new theory breaks C and T, leaving CT intact, and has to appeal to short-range forces to provide the observed breaking of P (Dirac 1973b). The calculations of Hari-Dass (1976) are therefore pertinent since by a consideration of the perihelion and gyroscope tests of general relativity he has been able to set a limit of 10^{-4} on possible C, P and T non-conservation in gravity. The expected level of symmetry breaking has not been calculated for Dirac's theory, but in P, at any rate, it is ambiguous because the relation between gravity and elementary particle theory is ill-defined at short ranges. The experimental investigation of gravity at short ranges has not been greatly pursued, one notable exception being the results of Long (1976), who claims to have found departures from an inverse square law at laboratory distances. Over scales of 4·5–30 cm, his experiments suggest that the gravity force can be represented by redefining G as distance-dependent above some standard value, G_0, such that $G(r) = G_0 (1 + 0.002 \ln r)$. This represents a short-range repulsion, and is a departure of the gravity law power from an exact value of two.

This departure can be compared with the analogous Coulomb law, where the exponent is exactly two within $\pm 3 \times 10^{-16}$, and sets an upper limit to the squared photon rest mass in geometrical units of $1 \cdot 04$ $(\pm 1 \cdot 2) \times 10^{-19}$ cm^{-2} (Williams *et al* 1971). It is convenient to note here that the possible non-equality of the charge between the electron and positron has been checked in an experiment using the flow of gas out of a special container by King (1962). He was testing a suggestion of Bondi and Lyttleton that 'neutral' hydrogen atoms might carry 2×10^{-18} units of an elementary charge as a net atomic charge, this being the basis of a cosmology lying outside general relativity (see King 1960, pp562–3). Chiu (1963, p241) has given the limit

$$\frac{e^+ - e^-}{e^+} \lesssim 10^{-22}, \qquad (3.25)$$

a limit which effectively disproves the prediction of Bondi and Lyttleton. Chiu also gives a limit for the velocity dependence of charge, on the hypothesis that

$$e^- = e_0^- + \frac{Av^2}{c^2}, \qquad (3.26)$$

in the form $A \lesssim 10^{-18}$ CGS. He further discusses the assertion of Mach's Principle that the inertial properties of matter are determined by the configuration of matter in the Universe. The anisotropy of mass caused by the presence of the Sun and the Galaxy might be expected to show up as a frequency shift in dipole transitions of the atoms of an apparatus rotated relative to the Sun or the Galaxy. No such effects are observed to an accuracy of one part in 10^{10}, a limit which tends to cast doubt on the validity of Mach's Principle. The Mössbauer effect and nuclear resonance of ^{6}Li atoms can, in fact, be used to establish the degree of isotropy to one part in 10^{22}, seriously embarrassing the relevance of Mach's assertion. Dicke (1961b) has extricated himself from this attack on the validity of Mach's Principle, on which principle the Brans–Dicke scalar–tensor theory of gravity is based (§3.3), by a not universally accepted

69

argument. He suggests that the anisotropy should be, in any case, unobservable, because all particles are expected to possess the same anisotropy.

Schiff (1960) has claimed that the relativistic redshift of light in a gravitational field and the deflection of light passing near the Sun can be obtained without using general relativity. The deflection of light calculation involves transformed special relativistic time intervals and the usual transformation of the radial length interval; one also needs a velocity of light that is anisotropic. He has concluded that only the perihelion advances of the planets really test general relativity. Schiff has, in reaching these conclusions, used the argument that since the gravitational redshift of light can be deduced from the Principle of Equivalence, there is no need for measurements to test the latter principle explicitly (the gravitational redshift of light has been observed to very high accuracy). Dicke, who has searched ardently for any signs of variation in the constants of nature, allowed by the limits attained in the Eötvös experiment and other tests, has naturally reacted against the claims of Schiff (Dicke 1953, 1960a, b). He has come to the conclusion that experiments do not, at the moment, rule out a small space variation of the fine-structure constant, α, of a size that might be expected to accompany an equivalent change in the gravitational constant G. Such variations, directly related to anisotropies of the inertial properties of matter are, in particular, predicted by certain metric theories of gravity.

3.7. Metric Theories and Anisotropy

The existence of gravitational anisotropies in the Solar System has been investigated by Will (1971a, b, d) in so far as metric theories of gravitation (such as the Brans–Dicke theory, with general relativity as a limiting case) predict anisotropies in the passive gravitational mass of rotating bodies such as the Earth and the Sun. A passive body with inertial mass, m_{I}, should accelerate in a Newtonian potential, ϕ^{p}, according to

$$m_{\mathrm{I}} a^{\alpha} = m^{\alpha\beta} \phi_{,\beta}{}^{\mathrm{p}} \tag{3.27}$$

where a^{α} is the acceleration and $m^{\alpha\beta}$ is the passive gravitational mass tensor, which is anisotropic in non-Einsteinian theories. This so-called Nordtvedt effect, to which other references are given by Will, causes a perturbation in the Earth–Moon distance. This is very small in size, and very hard to detect, as are the perturbations in the orbits of the planets due to the Sun's anisotropic passive gravitational mass. Metric theories (such as that of Whitehead) can also be formulated in a way that makes G anisotropic (Will 1971d). These theories can be tested by measuring G locally by a Cavendish experiment, and looking for anisotropies in the measured values of G as the Earth rotates. Gravimetric data used with this method rule out Whitehead's theory, since it predicts an effect 200 times larger than the observed limit. Metric theories such as this generally possess unequal values for the velocities of propagation of gravitational effects and light (although in the theory developed by Will (1971d) they are equal to within 2%). Will also gives a useful tabulation of limits on many unconventional, but feasible, accelerations, such as acceleration arising from a dependency on the velocity coupled with the gradient, for instance. Data like those tabulated by Will provide important restrictions on theories of gravity which predict anomalous behaviours of G.

Such theories are generally referred to as preferred-frame theories, since they only satisfy the three classical tests of general relativity in the mean rest frame of the Universe. (The connected relative-distance Machian theories have been discussed by Barbour 1974.) Since the Solar System has a velocity of about 200 km s^{-1} relative to the mean rest frame, certain observations can be carried out, in addition to the extant tests of general relativity, to test the theories of Page and Tupper, Yilmaz, Papapetrou, Ni, Coleman and Rosen (Nordtvedt and Will 1972). Several of these tests are geophysical, since the anisotropy of G would cause a periodic expansion and contraction of the Earth as it moves. This, although of very small amplitude, would cause a detectable change in the mean moment of inertia (I) of an amount $\Delta I/I \simeq -0{\cdot}1 \, \Delta G/G$ (cf the results of §3.4). Most possible such variations are larger than can be accepted, and place limits on the acceptability of theories giving rise to them. In a similar way, the possibility of a vector–tensor gravitational theory, as opposed to the scalar component of the Brans–Dicke theory, is restricted by the prediction of a 12 h sidereal solid-Earth tide and a yearly variation in the rate of rotation of the Earth.

A theory of this type, in which G as measured in a local system is unaffected by external masses, was first outlined by Will and Nordtvedt (1972), after they had demonstrated that all 'stratified' theories of gravitation must be incorrect. The name refers to those theories possessing a preferred rest frame that is conformally flat, with a metric depending on a scalar field, ψ_s, and a universal time coordinate, t, that can be put into the form

$$ds^2 = \exp\,[2f(\psi_s)]\,dt^2 - \exp\,[2g(\psi_s)](dx^2 + dy^2 + dz^2),$$

$$(3.28)$$

where f and g are two functions of ψ_s. The approach used by Nordtvedt and Will in dealing with theories that can be expressed in terms of this metric is to employ the parametrised post-Newtonian (PPN) formalism (Nordvedt 1968a, b, 1969, 1970, 1971, 1972, 1973, Will 1971a, c, Thorne and Will 1971). This approach confirms the unacceptability of the 1912 (non-general relativistic) theory of Einstein and the theory of Whitrow and Morduch. (The latter is equivalent to Einstein's theory with a variable velocity of light; see Will and Nordtvedt 1972.) A compendium of metric theories of gravity and their post-Newtonian limits has been given by Ni (1972), who defines metric theories to be those with a signature of -2 and an interval expressible as $ds^2 = g_{ij}\,dx^i\,dx^j$, and which satisfy $T^{ij}_{;j} = 0$, where T^{ij} is the energy-momentum tensor.

The Nordtvedt effect (which is really the breakdown of the Principle of Equivalence for massive bodies) puts close limits on all metric theories, as noted above for stratified theories. Nordtvedt (1968a, b) first pointed out that the Earth's gravitational-to-inertial mean mass ratio could differ from unity by an amount proportional to its internal energy. Such failures of the Principle of Equivalence for celestial bodies, because of their internal gravitational energy, would be observable as, for example, an anomalous oscillation in the lunar orbit in the case of the Earth–Moon system, that could be searched for using laser-ranging techniques (§6.7). Laser tracking could be employed to determine the several parametrised post-Newtonian coefficients of numerous metric theories of gravity, and so help to narrow down the field of possibilities. Ni considers that the only acceptable theories are general relativity and the Bergmann–Wagoner scalar–tensor theory with its special cases (that is, the Nordtvedt scalar–tensor theory and the Brans–Dicke–Jordan theory; the former is equivalent to the Bergmann–Wagoner theory with zero cosmological constant). Extensive work has been done by Ni on the Brans–Dicke–Jordan cosmology, but no reference is made to other cosmologies such as that of Hoyle.

The last omission has been rectified by Will (1972) who has set up a sort of acceptability algorithm to filter out theories of gravity that are incorrect. The theories of Hoyle (C-field) and Milne (kinematic relativity) are rejected as being incomplete; the theory of Poincaré and the two theories of Kustaanheimo are rejected as not being self-consistent, and the theories of Birkhoff and Yilmaz (1971 theory) are rejected as not agreeing with Newtonian physics to lowest order. Possibly, one might reject all non-metric theories on the grounds of the Eötvös experiment and the gravitational redshift of light. This last opinion, while widespread, I believe to be arguable, since the choice of geometrisation as a framework for a theory is no more fundamental than any other concept. This is supported by the work of Roxburgh and Tavakol (1975) who have shown that while theories like the Poincaré–Lorentz invariant theory were rejected on the ground of not being geometrisable in Riemannian space–time, this theory in particular is geometrisable in the more general Finsler space, and has cosmological solutions very similar to the Milne theory (§2.2). The Poincaré theory, while manifestly incorrect on the grounds of the basic general relativistic tests, is an example of a theory that looks non-geometrisable, but can, in fact, be geometrised given sufficient ingenuity. This is because there is a complete spectrum of possible formalisms, from total Machian geometrisation at one end to totally non-geometric theories, like Newtonian mechanics, at the other. There are theories in between that have an invariable background with a superposed variable geometry, caused by added masses. The Rosen bimetric theory (Rosen 1973, 1974) is of this type.

The Rosen theory is attractive in that the background metric can be thought of as the effect of a massive shell of matter located at infinity, the removal of this shell resulting in the recovery of general relativity. It seems to me that the Rosen background is equivalent to the vacuum field interpretation of the cosmological constant in Einstein's theory (see Chapter 7). While the equivalence of Λ and the Rosen invariable background has not been shown, the Rosen theory is rather dubious in status because of the difficulty of interpreting and defining the concept of mass in it. This problem also causes ambiguities in the theory's post-Newtonian formalism as compared with other theories. So far as other theories of the type being considered are concerned, the theories of Nordström Page and Tupper, Yilmaz (1958 theory), Papapetrou, Ni (two theories) and Coleman and Rosen can be rejected as not giving the correct perihelion advances of the planets. Whitehead's theory, as mentioned previously, can be rejected on geophysical grounds.

The only surviving theories at this stage are therefore general relativity, the Jordan–Brans–Dicke theory and the vector–tensor theory with the coupling parameter, k_V, such that $k_V < 3 \times 10^{-2}$ (Hellings and Nordtvedt 1973). The vector–tensor theory is attractive because it can be immediately connected with cosmology, in that k_V defines the gravitational constant by

$$G = \frac{G_0}{1 + \frac{1}{2}k_V^2} \tag{3.29}$$

and obeys the equation

$$\ddot{k}_0 + 3\dot{k}_0 \left(\frac{\dot{S}}{S}\right) - 3 \left(\frac{\dot{S}}{S}\right)^2 k_0 = 0. \tag{3.30}$$

Here, a zero subscript refers to conditions far from the local system, and $S = S(t)$ is the usual expansion factor defining the size of the observable Universe at any epoch t (a dot denotes differentiation with respect to t). While there are thus essentially only three surviving theories, it is quite feasible that some as yet undiscovered theory will also qualify as being potentially correct. Nordtvedt (1972) has reviewed the empirical status of gravitational theories as they affect Solar System experiments, and has given a metric expansion for any general metric field theory that becomes Minkowskian at infinity and is invariant under a Lorentz transformation. This expansion is in terms of a series in the velocity of a test system relative to some preferred inertial frame. In the general case, nonlinear terms in the products of the masses of the component bodies exist (representing the self-augmenting property of gravity as opposed to electromagnetism, for example), as well as terms equivalent to gravitational 'magnetic' effects.

To the theories that emerge from the acceptability algorithm (general relativity, Jordan–Brans–Dicke, vector–tensor) must be added that of Hoyle and Narlikar. The last reduces in many aspects to the Brans–Dicke theory, but the basis of the Hoyle–Narlikar cosmology is, in fact, more fundamental than that of the other mentioned metric theories. The basis of the theory lies in conformal invariance, and this can to some degree be taken over as a basis for any metric theory. As a concept, it is certainly necessary to consider it before embarking on an examination of the interrelationship of the physical constants. Conformal invariance itself is best approached by the related idea of kinematic equivalence.

3.8. Kinematic Equivalence

The arbitrariness of coordinate systems in general relativity, and the use of chosen frames for the sake of mathematical convenience in many metric theories of gravity, are realisations of what Milne called the study of equivalences. He considered that the consistent exchange of signals between a group of observers is the prime physical attribute of cosmology, to which both philosophical ideas about time (Whitrow 1961) and the practical working of clocks should both be subject. He took time as the fundamental quantity, and aimed to relegate distance to a supplementary, dependent role (Milne and Whitrow 1938). In a cloud of particles with observers (the equivalence), all reflecting light signals among themselves, it is possible for clocks on all the particles to be set into congruence. Once such a congruence is set up, it remains valid for any regraduation of the clocks from scale t to scale $T = \chi_M(t)$, provided $\chi_M(t)$ is monotonic. (T and t as used in this section have no connection with special epochs of the Universe as discussed elsewhere.) Mathematically, one special choice of

$\chi_M(t)$ stands out, namely, that which reduces all the other particles (that is, the other members of the equivalence) to rest with respect to one arbitrarily chosen particle. This is the famous τ time of Milne's cosmology. The other notable choice is t time, in which all the members of the equivalence are in uniform motion. The particles define a substratum, this being the equivalence in which the density distribution in space and time has the same *description* as viewed by every individual member. It is the origin of the idea of the Cosmological Principle. Particle motions in the substratum define a dynamics which is closely akin to that of conventional physics when referred to parametrisation in τ time, not t time.

In τ time, there are galactic redshifts, but they are due to secular changes in remote atoms as compared with nearby ones, these changes being interpretable as Doppler shifts in t time. If the galaxies define an equivalence, 'the Universe is its own clock', and τ time can provide a world-wide measure of simultaneity that is identifiable with Newtonian time (Milne and Whitrow 1938, p266). Only in τ time can two observers assign the same value to the epoch of an event, this being familiar physical time, locally. The possibility of parametrisations such as those defined by t and τ exists for accelerated as well as for uniformly moving observers, since the algebraic function that defines the equivalence is subject only to the constraint that the group concerned should contain a smallest member. Equivalences need not be linear. Observations of the galaxies suggest, however, that they do represent a linear equivalence in which light signals can be used to define a distance as $r \equiv c(t_3 - t_1)/2$, where a signal emitted at time t_1 has been sent out, reflected at another member of the equivalence, and received back at base at time t_3. The epoch of reflection is defined to be $t_r \equiv (t_3 + t_1)/2$. These are t time definitions. In τ time, *all* observers give the same epoch number to a distant event.

The precise form of the equivalence depends on the radar-mapping sequence originated by Milne. If the signal which left base A at t_1 was reflected at particle B at time t_2', as read at B, to be received back by A at t_3, then one can write

$$t_2' = \theta_M(t_1); \qquad t_3 = \phi_M(t_2') \qquad (3.31)$$

76

where θ_M and ϕ_M are certain functions to be found. Reciprocity implies that $\theta_M \equiv \phi_M$ be chosen. If B changes his scale from t' to T', where $T' = \chi_M(t')$ and $t' = \chi_M^{-1}(T')$, then

$$T_2' = \chi_M \theta_M(t_1); \qquad t_3 = \phi_M \chi_M^{-1}(T_2'). \qquad (3.32)$$

To obtain χ_M, one can solve $\chi_M \theta_M \chi_M^{-1} = \phi_M$ for χ_M, given θ_M and ϕ_M. Indeed, one can choose, in general,

$$\theta_M(t) = \psi_M \beta_M \psi_M^{-1}(t),$$

where β_M is a constant and ψ_M is a monotonic function defining the whole equivalence. The choice of ψ_M is left open. It is this arbitrariness that contains the basic factor in Milne's kinematic equivalence argument, and which has since been taken over in other forms by various metric theories of gravity under the guise of the irrelevance of the choices of coordinate and reference frames. The choice $\psi_M = t$ gives special relativity with uniform motion. The choice $\psi_M = \ln t$ gives the stationary equivalence. These are two cases of a general system, all of whose members are different descriptions, via differently graduated clocks, of the *same* kinematic events. The t and τ scales are related (Milne and Whitrow 1938, p284) by

$$\tau = t_0 \ln (t/t_0) + t_0. \qquad (3.33)$$

The concept of the equivalence makes no reference to the distribution of the particles which are its members; in this respect the principles of indistinguishability, uniformity and reciprocity of description on which it is based are more general than what has come to be called the Cosmological Principle. The equivalence as defined so far admits not only uniform-density models but also, for example, inhomogeneous models, provided they satisfy the requisites of forming a substratum.

In t time, the distribution of particles away from any chosen observer is not uniform, but goes as

$$n_M(t, r) = \frac{B_M t}{c^3(t^2 - r^2/c^2)^2} \qquad (3.34)$$

where B_M is a constant. There is still no preferred observer, even though the distribution is non-uniform, increasing away from a local origin. In τ time, the density of the stationary particles is

$$n_M = \frac{B_M}{c^3 t_0^3} = \text{constant.} \qquad (3.35)$$

While $n_M \simeq n_M^0(1 + 4r/ct)$ nearby in t time, the apparent singling out of the local observer as preferred is an illusion. In t time, photons have constant frequencies and atoms have constant energies; while in τ time, gravitational systems have a very simple description but the behaviour of photons is complex. Milne and Whitrow believed that photons keep t time and that dynamics keeps τ time. The two scales become out of step to an increasing degree for more remotely observed sources, the discrepancy being synonymous with the redshifts of the galaxies. Thus, τ time gives dynamics its simplest form, and t time gives optics its simplest form. The view of Milne and Whitrow was that general relativity as usually employed is unnecessarily restrictive in employing the observer's 'proper time', whereas a choice of τ time, in which there is a kind of simultaneity, is also allowable.

The basic importance of the Milne equivalence for cosmology, in addition to its role in stimulating research into the arbitrariness of reference frames, lies in the way in which it shows how an apparently non-uniform background of particles can be interpreted as a symmetric substratum, provided the equivalence is regraduated. The problem, if this hypothesis is accepted, is to delineate which clock graduation is relevant to the large-scale Universe, since it now seems very unlikely that a density distribution of the form of equation (3.34) can be reconciled with observation. A large body of opinion favours a uniform background, in which case equation (3.35) holds, and τ time would seem to be involved, bringing in some rather peculiar dynamical consequences. It seems to me that there may exist a third natural function, $\psi_M = \Gamma(t)$, that defines a clock graduation in which Γ is a scale of time suited to the parametrisation of a hierarchical distribution of matter. (The school believing in a hierarchical distribution of galaxies and clusters will be considered briefly in Chapter 7.) The discovery of the exact relation between Γ and the conventional time scale of physics might be a difficult task. Irrespective of its connection with the density distribution of the Universe, the idea of the equivalence has been of great influence in inspiring, and giving a basis to, the now more widely employed concept of conformal invariance.

3.9. Conformal Invariance

That some metric theories were conformally invariant and that a class of metrics in general relativity possessed conformal properties had been known for a long time before its importance for astrophysics and electrodynamics was realised. The main contributors were Hoyle and Narlikar (1974), although the arbitrariness that its application uncovers in cosmology seems to have been comprehended even by Milne, as will be seen below.

If a space–time manifold, M^g, in which the metric is g_{ij}, is transformed to a new space–time, M^{g*}, with a metric $g_{ij}*$, in such a way that

$$g_{ij}* = \Omega^2 g_{ij} \qquad (3.36)$$

then the transformation is said to be conformal provided that Ω is a finite, nonzero function of the coordinates. Elements of four-dimensional length transform similarly as

$$ds* = \Omega \, ds. \qquad (3.37)$$

The fundamental importance of conformal transformations lies in the fact that the ratio of two lengths, possibly in different directions but at the same point, does not change under the transformation. Any quantity is conformally invariant if it does not change under equation (3.36). The electromagnetic action, A_i, leading to Maxwell's equations of electrodynamics is conformally invariant, as is the Weyl conformal curvature tensor, but Einstein's equations are not (Hoyle and Narlikar 1974, p28). The space–time M^g is further said to be conformally *flat* if a coordinate system exists in which

$$g_{ij}* = \eta_{ij}; \qquad g_{ij} = \Omega^{-2} \eta_{ij}, \qquad (3.38)$$

where η_{ij} is the Minkowski metric. The Robertson–Walker $(\Lambda \equiv 0)$ spaces are conformally flat. For example, the steady state Universe (Bondi and Gold 1948, Hoyle 1948) has a metric commonly written with an epoch coordinate t, as

$$ds^2 = c^2\, dt^2 - \exp(2Ht)[dr^2 + r^2(d\theta^2 + \sin^2\theta\, d\phi^2)], \quad (3.39)$$

which can be written in the conformally flat form

$$ds^2 = (1/Ht')^2[c^2\, dt^2 - dr^2 - r^2(d\theta^2 + \sin^2\theta\, d\phi^2)], \quad (3.40)$$

with t' as time coordinate. Transformations for the $k=0$ Einstein–de Sitter model, the $k = +1$ closed model, and the $k = -1$ open model are given by Hoyle and Narlikar (1974, pp126, 130 and 134, respectively). The cosmological constant Λ is usually taken as zero in conformally flat Friedmann cosmologies (Hoyle 1974, p331), but presumably there are other inhomogeneous models with finite Λ that satisfy the criterion.

It is advocated by Hoyle and Narlikar that the physical laws, besides being invariant in the general relativistic sense under coordinate transformations, should also be invariant under conformal transformations of the geometry. This is connected with the meaning of dimensionality, and the desire to remove redundancies from our description of physical systems. Consider, in the first place, quantum mechanics. The equations of motion can be obtained from an action

$$S_A = -\int_{x_1}^{x_2} m\, dx + e\int_{x_1}^{x_2} A_i\, dx^i$$

so that a wavefunction can be formed (Hoyle 1974, p302) by summing over all possible paths:

$$\psi = \sum_{\text{paths}} \exp(iS_A/\hbar).$$

Here the $\hbar$ factor is inserted to make S_A dimensionless, but $S_A = [0]$ can be arranged in the first place by defining a suitable Lagrangian. (Square brackets will be used to denote dimensionality.) In the non-relativistic theory, the Lagrangian can be taken to be

$$\mathscr{L}_A = \frac{m\dot{x}^2}{2\hbar} - \frac{eA_i}{\hbar}\frac{dx^i}{dt}, \quad (3.41)$$

where m is the particle mass. By classical field theory, A_i has dimensionality eL^{-1}, where e is the elementary charge, so that

80

$A_i = [eL^{-1}]$ and the last term in equation (3.41) is $eA_i\hbar^{-1}$ $(dx^i/dt) = [e^2\hbar L^{-1}]$. Since quantum mechanics can be derived from an action, S_A, in which $S_A/\hbar$ must be a dimensionless phase angle, clearly $\mathscr{L}_A = [L^{-1}]$. Therefore, with this way of writing things, $e^2/\hbar = [0]$ is the fine-structure constant with an observed value of $7 \cdot 297351 \times 10^{-3}$. The properties of physical systems are determined by $\mathscr{L}_A$, and e only ever appears in association with $\hbar$ in physics as $e^2/\hbar$. Hence, $\hbar$ can be absorbed into e^2, making the dimension system less redundant and fixing $e^2 = 7 \cdot 297351 \times 10^{-3}$. A similar approach can be made in the relativistic theory, in which the Dirac equation for a free particle is

$$\gamma_D{}^k \frac{\partial \psi_D}{\partial x^k} + \frac{im}{\hbar} \psi_D = 0, \tag{3.42}$$

where $\gamma_D{}^k$ are the Dirac matrices ($k = 1 \rightarrow 4$). The mass only occurs as $m/\hbar$, so by absorbing $\hbar$ one fixes $m = [L^{-1}]$. All physical quantities, by removing redundancies of this kind, can be expressed in terms of a length and the unit of elementary charge. The standard length is chosen to be the Compton wavelength of the electron, so that

$$C = m_e^{-1} = [L],$$

and $(e^4/4\pi)C^{-1}$ is the Rydberg constant, giving the particle separation as $e^{-2}C$ in macroscopic bodies.

All of physics except gravitation is concerned with the three coupling constants for the strong, weak and electromagnetic interactions, the last being the fine-structure constant, now defined as e^2. In conventional units $2\pi C = 2 \cdot 4263096 \times 10^{-12}$ m. In the actual Universe, measurements are made on bodies with redshifts in their spectra defined by $(1 + z) \equiv \lambda(\text{received})/\lambda(\text{laboratory})$. Finite z values could either be due to motion of the sources, a non-flat space–time ($g_{ij} \neq \eta_{ij}$) or to a spatial variation of C (so that m_e is a function of the position X). There is no way in practice to prove that C is the same at all X, so Hoyle and Narlikar admit the possibility that $C = C(X) = m_e^{-1}(X)$. Variations in the metric can be expressed in terms of this concept and vice versa, because the same electromagnetic wave phase factor at X is produced by the same set of paths with A_i^* in M^{g*} as by A_i in M^g, by virtue of the conformal invariance of the Maxwellian action, A_i $(A_i^* = A_i)$. Cosmological problems can, in view of the conformal invariance of Maxwell's equations, be solved in a flat space–time η_{ij} with $C = C(X)$, just as well as by adopting the conventional usage of a curved g_{ij}. While local Doppler redshifts are accepted, in the Minkowski frame the overall expansion motion of the Universe is reduced to rest by choice, since there is no way physically to identify the conventional expanding motion of the galaxies as they are observed by electromagnetic radiation. The $C = C(X)$ dependency is open to observational investigation.

The impossibility of determining the physical laws with respect to some particular geometry results in the postulate that all such laws should be invariant under conformal transformations of the geometry (these are then physically equivalent). Conventional particle dynamics is not conformally invariant classically, and the relativistic theory is only so if the Dirac equation is made conformally invariant by choosing

$$m^* = \Omega^{-1}m; \qquad \psi_D^* = \Omega^{-3/2}\psi_D. \qquad (3.43)$$

A suitable particle dynamics on the nuclear level is lacking, although Hoyle is of the opinion that the Hoyle–Narlikar lepton theory (e, μ) might be a first approximation to an acceptable theory of the Salam–Pati type, based on a very large symmetry group (in this case SU32). The particle dynamics in the electromagnetic domain is chosen by Hoyle and Narlikar to be a direct-particle field, in which the potentials giving rise to the fields are defined in terms of the particle world lines and have no degrees of freedom of their own. There is a kind of action at zero four-dimensional distance, and the classical self-action problem is avoided. The choice of a direct electromagnetic-particle field is connected with the desire to obtain the empirically observed electromagnetic arrow of time within a theory that makes no *ad hoc* assumptions about the admissability of retarded and advanced solutions. The Wheeler–Feynman absorber theory is therefore demanded (see Hoyle and Narlikar 1964a), and cosmological models should be consistent with the perfect absorber hypothesis.

This theory concerns a selection of advanced and retarded

fields acting on a collection of charged particles confined to a local region of space–time. If $F^{ik}(X)$ is the usual antisymmetric second-rank field strength tensor formed by combining the magnetic and electric field strengths (or defined in terms of the four-potential by $F^{ik} = A^{i,\,k} = A^{k,\,i}$, where a comma denotes partial differentiation), then the Maxwell–Lorentz equations give the field acting on one chosen charge (i) due to all the others (j) as

$$\sum_{j \neq i} F_{\text{ret}}{}^{(j)} + \tfrac{1}{2}[F_{\text{ret}}{}^{(i)} - F_{\text{adv}}{}^{(i)}] + F_{\text{in}}. \qquad (3.44)$$

Here, F_{ret} is the retarded field of a single charge, which, acting on itself, can be decomposed into a source-free part $(F_{\text{ret}} - F_{\text{adv}})/2$ giving rise to the observed finite radiation damping force, and a part with source in the particle of $(F_{\text{ret}} + F_{\text{adv}})/2$ which contributes to the self-energy of the particle. The part F_{in} is taken to refer to the source-free fields coming in from infinity. There exists a second solution to the equations that is symmetric in time, namely

$$\sum_{j \neq i} F_{\text{adv}}{}^{(j)} - \tfrac{1}{2}[F_{\text{ret}}{}^{(i)} - F_{\text{adv}}{}^{(i)}] + F_{\text{out}}, \qquad (3.45)$$

where F_{out} represents the outgoing field just as F_{in} represents the possible incoming field, the net radiation field being $F_{\text{rad}} = F_{\text{out}} - F_{\text{in}}$. The average of equations (3.44) and (3.45) gives another time-symmetric form:

$$\tfrac{1}{2} \sum_{j \neq i} [F_{\text{ret}}{}^{(j)} + F_{\text{adv}}{}^{(j)}] + \tfrac{1}{2}[F_{\text{in}} + F_{\text{out}}]. \qquad (3.46)$$

The conventional Sommerfeld radiation condition, $F_{\text{in}} = 0$, gives the usual fully retarded solutions that seem to be observed, simultaneously destroying the basic time symmetry of the theory. In the absorber theory, the basic time symmetry is preserved by choosing the boundary condition $(F_{\text{in}} + F_{\text{out}}) = 0$, so that every charged particle generates retarded and advanced fields. The response of the Universe is such that fields appear to propagate in the familiar manner, provided there is perfect absorption along the future light cone.

The absorber theory has been well reviewed by Davies (1974c, pp130–52), who points out that the existence of advanced and retarded solutions to the electrodynamical equations, with the prediction of an absorbing Universe as a consequence, is basically a phenomenon open to test. Only a few of the commonly discussed models satisfy the absorber criterion (Davies 1974c, p149, Hoyle and Narlikar 1974, p47), among which are the steady state model with metric equation (3.39), the Minkowski model and the $k = +1$ Friedmann model. (Landsberg and Park (1975) have examined the evolution of entropy from cycle to cycle of a closed Friedmann model with gas–dust–radiation interaction.) Of these, only the first gives an unambiguous retarded electromagnetic propagation, as observed. It should, however, be mentioned that an alternative self-consistent solution of the absorber theory exists, in which there are fully retarded fields acting on any chosen particle, although the future null cone is transparent. There could be other models satisfying this solution that are of the $k = 0$ and $k = -1$ types.

Hoyle and Narlikar (1974) have considered several models satisfying the conformal invariance and electromagnetic response conditions; the latter, in particular, providing support for the existence of both positive and negative contributions to the mass field that manifests itself in $m_e(X)$. The particle masses are related to the propagator, ϵ_{HN} introduced in §2.7; if two particles (a and b) are pursuing separate paths, but interact via, for example, the exchange of a boson, the interaction is characterised by ϵ_{HN}. The action, defined by

$$S_A = -\sum_a \int m_a \, da,$$

can be varied with $\delta S_A = 0$ to first order, the metric being varied from g_{ij} to $g_{ij} + \delta g_{ij}$. The resulting field equations (Hoyle 1974, p313) for a particle of mass m give

$$Gm^2 = \frac{3\,\epsilon_{HN}{}^2}{4\pi}, \tag{3.47}$$

as mentioned in §2.7. The sign of G, manifested as the always attractive property of gravity, is correctly predicted, even though the sign of ϵ_{HN} is not fixed. The equations of motion can also be obtained from the action (Hoyle 1974, p316), or by taking the divergence of the field equations. Observationally (via G), $\epsilon_{HN}^2 \approx 10^{-40}$ and is dimensionless, implying that particles are very effectively self-shielded in interactions ($\epsilon_{HN}^2 \approx 1$ would correspond to an elementary mass quantum of order 10^{16} g).

If $\epsilon_{HN} = \epsilon_{HN}(X)$ is admitted as in the Hoyle–Narlikar theory discussed in Chapter 2, then G becomes a function of the coordinates, the precise dependence being open, but allowing, for example, the Dirac behaviour $G \propto t^{-1}$. The G behaviour of the most natural model is rather modest, being $G \propto t^{-1/2}$, but this model appears to conflict with the classical tests of general relativity (Hoyle and Narlikar 1974, p190) unless a response condition for the mass field equivalent to that for the electromagnetic field is employed. This provides indirect evidence in favour of the Wheeler–Feynman absorber theory. The size of ϵ_{HN} could only be predicted from a suitable dynamics of elementary particles, as noted above. Since there is no reason to deny the possibility that ϵ_{HN} can be negative as well as positive, there must be naturally occurring surfaces of $\epsilon_{HN} = 0$ between domains where it has opposite signs. Hoyle assumes that all the $\epsilon_{HN} = 0$ surfaces are closed, although it does not seem that this need necessarily be so. The surface of $\epsilon_{HN} = 0$ in the conformal Minkowski frame, in which the Hoyle–Narlikar theory is developed, corresponds to the big bang event in the expanding Robertson–Walker frame. Near a surface of $\epsilon_{HN} = 0$, it is a good approximation to use a homogeneous Robertson–Walker model, the present epoch of the Universe being such as to allow this. Surfaces of zero ϵ_{HN} on a small scale must alter the gravitational collapse picture, since collapse to a black hole in a zone of $\epsilon_{HN} > 0$ corresponds to the emergence of a white hole on the other side of the surface, where $\epsilon_{HN} < 0$, so that in a way matter can emerge from a black hole.

On the large scale, the existence of another region on the opposite side of the $\epsilon_{HN} = 0$ surface results in some unusual astrophysical results, notably in connection with the 3 K microwave background. On the Hoyle–Narlikar picture, the black-body field represents thermalised starlight that was originally generated on the other side of the $\epsilon_{HN} = 0$ surface (Hoyle 1975). The thermalisation occurred because the

Thompson cross section (e^2/m_e) diverged near $\epsilon_{HN} = 0$ (identified as $t = 0$), since $m_e \simeq 0$ at $t \simeq 0$. Davies (1975) has queried this result on the ground that irreversible processes such as those noted might have been expected to have increased the entropy of the Universe without limit, compared with the observed finite value as represented by the 3 K background. There are certainly some curious effects attached to passage through the $\epsilon_{HN} = 0$ surface. Near $t = 0$, galaxies were crowded together, and Rees and others have raised queries concerning astrophysical processes at such epochs. Particle masses vary with time in the Minkowski frame, of course, the dependency being calculable for the Friedmann models simply by obtaining Ω from the metric and the field equations and then using $m^* = \Omega^{-1}m$ as in equation (3.43).

In the general case, the dependencies involved are not known *a priori*, but remain to be found as acceptable solutions to the field equations (Einstein's equations in the conventional frame with G and particle masses being constant). All the non-gravitational physics of these solutions, in summary, can be characterised by (i) three dimensionless coupling constants for the strong, weak and electromagnetic interactions; (ii) the particle mass ratios, and (iii) the mass of some chosen reference particle. The gravitational physics comes in as a kind of criterion system for acceptable model universes based on the elimination of redundancies in dimensional arguments, the principle of conformal invariance and the perfect absorber theory of radiation. Of these, the need for conformal invariance would seem to agree with our intuitive conception of space–time. One could argue that if it should turn out that the principle is incorrect, then consideration would have to be given to explaining why it is incorrect.

While the Hoyle–Narlikar cosmology makes good use of conformal invariance, it would seem likely that a definitive theory of this type should provide a straightforward explanation of the origin of inertia and the expected inertial anisotropy of matter (§§3.2 and 3.3). The latter might, perhaps, be formulated in terms of a space-variable value of G (§3.4). The absolute value of this parameter might be connected with the physical nature of the vacuum (§3.5). Any metric theory of gravity of this type should survive the selection algorithm discussed in §3.7, and conform to the results of the Eötvös experiment (§3.6). The equivalence concept (§3.8) may be instrumental in indicating which theories are acceptable by identifying the most relevant cosmological coordinate frames for the observable Universe.

4. The Eddington Numbers

4.1. Introduction

Certain numbers, whether naturally occurring or algebraic, have always held a fascination for intellects of an inquiring disposition. This is manifestly true of a special class of dimensionless numbers that appear in physics and cosmology, which have been the origin of much fundamental research and an even larger amount of more questionable speculation. It is now usual to refer to these numbers as 'Eddington numbers', although the term has come to have a wider significance than it meant to Eddington. He was, however, the prime investigator to draw attention to the peculiarity of some of the parameters appearing in cosmological calculations.

The dimensionless numbers concerned have been mentioned in previous chapters as basic attributes of certain cosmologies that go beyond general relativity. It is convenient, however, to state explicitly the four main dimensionless constants in nature (Bertotti *et al* 1962, p10) which have quite distinctive magnitudes.

(1) The coupling constant for strong interactions is $\simeq 10$. The fine-structure constant for electromagnetic interactions is $\alpha \simeq 1/137$. On a logarithmic scale, both are of order $\approx 10^0$.
(2) The weak coupling constant is $g_w = (g_\beta m_\pi{}^2 c/\hbar^3)^2 \simeq 3 \times 10^{-14}$, where g_β is the β decay coupling constant, $g_\beta = 1 \cdot 4 \times 10^{-49}$ erg cm^3 and m_π is the pion mass (c is the velocity of light and $\hbar$ is Planck's constant$/2\pi$).
(3a) The ratio of gravitational to electrical force in the hydrogen atom is $Gm_e m_p/e^2 \simeq 5 \times 10^{-40}$ (where m_e is the electron mass, m_p is the proton mass, and e is the electron charge); which is approximately equal to (3b) the ratio of an atomic period to the age of the Universe, $e^2/m_e c^3 T \approx 10^{-40}$, and this is

88

of the same order as (3c) the dimensionless gravitational coupling constant $Gm_p^2/\hbar c \simeq 5 \times 10^{-39}$, which is approximately equal to (3d) the reciprocal of the number $Tm_e c^2/\hbar \approx 10^{40}$.
(4) There is some reason for thinking that the number of particles in the Universe is $4\pi R_U^3 \rho_U/3m_p \approx 10^{80}$ (where R_U = the Hubble radius of the Universe $= cT \approx 10^{28}$ cm, and $\rho_U \approx 10^{-30}$ g cm^{-3}), so that the reciprocal of this number is $\approx 10^{-80}$.

There is clearly some arbitrariness present in the exact formulation of these numbers. The ones I have noted are those originally used by Dirac and discussed by Bertotti *et al* (1962). Often they are written with m_p replacing m_e, as for example in Dicke's discussion of Dirac's hypothesis (Dicke 1961a). The numbers are of such sizes that the question of whether m_e or m_p should appear is not important numerically. The main point is the appearance and progression of the numbers (inverses of some of the above) of sizes 10^0, 10^{40} and 10^{80}.

Some of the numbers are connected with cosmology, some with atomic and nuclear physics, and in particular, the coupling constants for the four basic forces in nature are involved. These forces, the strong, electromagnetic, weak and gravitational forces, have strengths in the ratios $1:1/137:10^{-13}:10^{-39}$, and are clearly related to the Eddington numbers, although there are numerous theories of them that do not involve cosmology, such as the unified gauge theories of elementary particles (Leader and Williams 1975; §2.8). The cosmologies previously treated have connections with the numbers (1)–(4), as we have noted, but the dimensionless Eddington numbers have generated so much comment since they were first promulgated that it may be relevant to consider briefly the evolution of ideas attendant on them as they concern gravitational theories.

4.2. Theoretical Background

Eddington thought that some of the numbers (1)–(4) were implicitly connected with the solution of certain elementary types of algebraic equation (Eddington 1929, 1935, 1939). He aimed to derive all the fundamental constants of physics from algebraic manipulations based on *a priori* ratiocination and not including appeal to empirical data. Some of his results appear at first sight to be in amazingly good agreement with observation, but the agreement is, in fact, *too* good. Jeffreys analysed the errors attendant on Eddington's calculated values as compared with observation, and found a suspiciously high level of exactitude. This led him to believe that Eddington had unwittingly included some empirical material in his analysis (Jeffreys 1967, p310). The fault seems to me to lie in the philosophy of the problem. It is possible to think of numerous relationships from which numerical results of arbitrarily close approach to a required figure can be obtained, but such an occupation is of minimal practical value. Whether or not this is what Eddington did it is difficult to say, because most subsequent investigators who have examined Eddington's work on his later development of the theory have come away baffled by the turgid nature of its argument. This is especially true of his book *Fundamental Theory* (Eddington 1946) and has been commented on by several workers who have sought to uncover its philosophical and mathematical bases (Dingle 1954, Slater 1957). Attempts to replace defective parts of the statistical aspect of the theory by new arguments (Kilmister and Tupper 1962) have been made, while interest in the whole theory continues to exist, despite the heavy criticism to which it has been subjected.

While I do not consider obscurity to be an adequate excuse for dismissing Eddington's approach, I do think that the part of it concerning the dimensionless numbers can be severely criticised on the following grounds. The constants of physics deduced by Eddington are supposed to comprise the essential data of nature, yet to claim that these numbers can be obtained from a theory leaves no place for the possible future discovery of new physical constants. The set of physical constants available at any epoch in the history of science will not be complete unless science has reached a dead end at that epoch. On the contrary, any theory aiming to explain the significance of the numbers (1)–(4) in a deductive manner should be open-ended to allow for the incorporation of the results of yet undiscovered laws of physics. This view is obviously justified in retrospect in the case of Eddington's work, which was formulated before the enormous widening of modern research in elementary particle physics. Despite this seemingly self-evident need for an open-ended approach to a possible theory of the numbers (1)–(4), a willingness to incorporate self-redundancy into such theories has been, and still is, lacking.

The failing of being 'over-complete' also applies to the cosmology of Milne, whose theory of world structure (Chapter 2) left no grounds for extension, although one cannot but admire the way in which intractable aspects of the theory were developed by ingenious and devious methods that have since found wider application. While Milne discussed the concept of dimensionality as it involved G, Dirac was the first to formulate a hypothesis of its possible time variability that was easily understandable. The large-numbers coincidence provided, at the time, only a crude cosmology, but Jordan and other workers were sufficiently impressed to attempt the construction of field theories of gravity that involved variability of physical parameters such as G. Brill has given an excellent review of the origins and formalism of the theory of Jordan, which was the first notably successful attempt to provide a secure theoretical foundation for Dirac's hypothesis (Brill 1962). Other attempts at the time due to Thirry (see Brill 1962, p51) and Fierz (see Fierz 1956, Jordan 1959) were similar to Jordan's, but less complete. The Jordan theory attracted a great deal of attention for a period, firstly because of its prediction that new matter can be created out of empty space (some of this appearing as already formed stars) and, secondly, because of its variable-G aspect. The formulation of the Jordan theory in which G varies results because the gravitational potential field surrounding a particle moving slowly on a geodesic in nearly Minkowskian space is slightly different from the normal form. The difference is most conveniently expressed in a renormalisation of the gravitational constant

$$G = G_0 \left(1 - \frac{1}{2B}\right) \tag{4.1}$$

where B is a parameter connected with the energy-momentum tensor of the source and G_0 is a constant (Brill 1962, p59; cf §3.4). Such renormalisation to account for effects of the energy-momentum tensor, T_{ij}, is common to all metric theories of the type proposed in connection with the Eddington numbers.

In general relativity, the field equations can be written as

$$\frac{c^4}{8\pi G}\left(R_{ij}-\frac{R}{2}g_{ij}\right)=T_{ij} \tag{4.2}$$

with the right-hand side being the forcing term (trying to curve space–time) and the left-hand side the restoring term, like the elasticity interpretation of Sakharov (§3.5). Any modifications to the equilibrium system, such as quantum corrections, modify equation (4.2) in the manner noted by Zeldovitch (1974) to

$$\frac{c^4}{8\pi G}\left(R_{ij}-\frac{R}{2}g_{ij}\right)+B'\left(R_{ij}-\frac{R}{2}g_{ij}\right)+B''\left(R_{ijlm}R^{lm}+\ldots\right)$$

$$+\ldots+Cg_{ij}=T_{ij}. \tag{4.3}$$

The Cg_{ij} term represents the effect of the cosmological constant, while the *observed* gravitational constant is given by equation (4.1) with $B=c^4/16\pi G_0B'$.

Whereas G does not vary with time in general relativity, in Jordan's theory G is time-dependent mainly because the Universe is evolving and is not asymptotically flat, the matter creation in the space–time being introduced as a statistical theory (Brill 1962, p64). Near large masses, space is severely curved, and it is interesting to note that Thorne and Dykla (1971), following a suggestion by Penrose (1970), have shown that black holes in the Jordan–Brans–Dicke theory are identical to those in general relativity. The Jordan-type theory thus reduces to Einstein's theory in the limit of extreme fields, but can differ sensibly from it in the domains of applicability to geophysics and non-extreme astrophysics. The geophysical consequences of the form of Dirac's hypothesis outlined above have been examined by Jordan (1962b), but to all intents the Jordan theory was superseded by that of Brans and Dicke. The latter cosmology has been extensively developed by Dicke both along geophysical and dimensional lines. While previous theories had been preoccupied with the meaning of G, Dicke widened the field of enquiry to include the other physical constants, also refocusing attention on the question of coordinate and space–time transformations, which had been the cornerstone of Milne's theory.

The Hoyle–Narlikar cosmology, which appeared subsequently to that of Brans and Dicke, reflected an understanding of the significance of coordinate and dimensional transformations that had taken several decades to mature. The full theory described in §§2.7 and 3.9 was thus the end product of a series of formulations which it is convenient to comment on here. The steady state theory was important in guiding this evolution of ideas, both in its original form, followed by the C-field, which was eventually interpreted as a direct-particle field (Hoyle and Narlikar 1964b). It also gave rise to interest in the gravitational influence of direct-particle fields (Hoyle and Narlikar 1964c) plus a new theory of gravitation (Hoyle and Narlikar 1964d), as described elsewhere. The new theory resulted in certain effects caused by the non-conservation of baryons (Hoyle and Narlikar 1966a) and ended in a radical departure from the steady state concept (Hoyle and Narlikar 1966b, c). The eventually emerging gravitational theory of the cosmology was a conformal theory (Hoyle and Narlikar 1966d). Conformal invariance in physics and cosmology has been summarised by Hoyle and Narlikar (1972c), while Narlikar (1974a, b) has provided an action-at-a-distance electrodynamics to complement the similar gravitational theory, the electrodynamics being an absorber theory that can be put into conformity with general relativity. A review of the evolution of the whole theory has been given by Hoyle (1974, pp333–43), together with an evaluation of the changing status of the steady state theory as judged on astrophysical grounds.

The observational validation of cosmological models is a difficult process (Davidson and Narlikar 1966), but it gradually became clear that the steady state theory in its traditional form could not be correct, and the new theory now stands largely emancipated from its predecessor. Both the cosmological and quantum aspects of the theory, interpreted from the viewpoint of action at a distance (four-dimensional distance), have been summarised by Hoyle and Narlikar (1974). There can be little doubt that the Hoyle–Narlikar formalism, with the cosmological and conformal parts taken together, represents a very logical theory of gravity and astronomy. This does not mean that it is in any sense better than Einstein's theory, but that the latter, if it is to continue to be accepted as the basic reference theory, ought to be put into a canonical form incorporating a proper account of conformal transformations. The Hoyle–Narlikar cosmology, besides being connected with G variability, is directly involved with those attempts, to be examined below, in which the Eddington numbers are explained by hypotheses directly implicating the observer. Notably, the carbon/oxygen ratio in the world is a crucial factor in providing the conditions necessary to allow observers such as human beings to live. This ratio depends on mechanisms of nuclear and atomic physics that are affected, via the coupling constant of the theory, by cosmology. Besides incorporating a proper account of dimensional transformations, it was also realised by Hoyle that this theory should not be closed in the sense discussed previously; new theories must develop as human minds evolve, hopefully in a way such that the new theories supersede, but encompass the old ones as time progresses. This process will probably not converge. It may be that one way to infer new knowledge about possibly acceptable cosmological formalisms lies in a proper understanding of the Eddington dimensionless numbers.

4.3. Stellar Astrophysics

It has been seen in Chapter 2 that Dirac attempted to account for the numbers (3) and (4) in terms of a relationship involving the evolution of the Universe. Dicke (1961a) has re-examined this argument. Since the age of the Universe ($T \approx 10^{10}$ yr) is peculiar to the present epoch, Dicke considered it unlikely that the numbers (3c), (3d) and (4) vary with time, as in Dirac's theory

$$\frac{Gm_{\mathrm{p}}^2}{\hbar c} \propto t^{-1}; \qquad \frac{Tm_{\mathrm{p}}c^2}{\hbar} \propto t; \qquad \frac{M_{\mathrm{U}}}{m_{\mathrm{p}}} \propto t^{+2}. \tag{4.4}$$

If they did, the preserved interconnection between the numbers would itself have to be independent of time, with the coincidence in sizes holding only at the present epoch, T. Dicke suspected that the first and last of equation (4.4) are true invariants, and that the second, while varying like t, is related

94

to the existence of observers (human beings) in the Universe at the present time. This argument rests on two bases. The first is that the maximum lifetime of a star of mass M_* (with emission and opacity given by free–free transitions, that is, bremsstrahlung) is

$$T_{\max} \approx \left(\frac{m_{\mathrm{p}}}{m_{\mathrm{e}}}\right)^{5/2} \left(\frac{e^2}{\hbar c}\right)^3 \left(\frac{Gm_{\mathrm{p}}^2}{\hbar c}\right)^{-1} \frac{\hbar}{m_{\mathrm{p}} c^2}. \tag{4.5}$$

Dropping factors (like 10^{-3}, etc), which are of order of magnitude unity compared with 10^{40}, gives

$$\frac{m_{\mathrm{p}} c^2 T_{\max}}{\hbar} \approx \left(\frac{Gm_{\mathrm{p}}^2}{\hbar c}\right)^{-1}, \tag{4.6}$$

agreeing with the rough equality of the numbers $(3d)$ and $(3c)$. The mass of the star in this analysis has a lower bound fixed by the assumption that electron degeneracy occurs at the appropriate nuclear reaction temperature. This gives

$$\frac{M_*}{m_{\mathrm{p}}} \simeq 10^{-3} \left(\frac{\hbar c}{Gm_{\mathrm{p}}^2}\right)^{3/2} \simeq (10^{40})^{3/2}, \tag{4.7}$$

a number that is commented on again in following sections. The minimum lifetime of a star, fixed by stability analyses, is of order $T_{\max}$, and both are of order T. One partial explanation of the Eddington numbers could be arrived at by realising that human beings, since they live near a star, only exist to ratiocinate on the numbers (3) because a time $t \approx T$ has elapsed, this being the probable age of a star like the Sun. The second basis of Dicke's argument lies in the postulate that the mass, M_{U}, of the Universe is related to its size, $R_{\mathrm{U}} = cT$, and age by

$$\frac{GM_{\mathrm{U}}}{c^3 T} \simeq 1, \tag{4.8}$$

this being a possible manifestation of Mach's Principle, which also explains the smallness of $(3c)$ as a result of the large amount of matter in the Universe. Since the combination of equations (4.6) and (4.8) gives an equation equivalent to the combination of numbers $(3c)$ and (4), one can say in sum that Mach's Principle, or another hypothesis giving rise to equation (4.8), plus the existence of observers at the present epoch, provides a possible explanation of the sizes of the numbers (3c), (3d) and (4), in which only the second is time-variable.

4.4. Anthropomorphism

The point of view advocated by Dicke as described above has been extended somewhat and put on a slightly different footing by Carter (1974), who has given explanations of some of the Eddington numbers based on what are referred to as the 'weak' and 'strong' anthropomorphic principles. The weak principle, as used by Dicke, means that we, as observers, must be prepared to allow that our location in the Universe is necessarily privileged to *some* extent, being dependent on our existence as observers. The strong principle states that the Universe must be such as it is to admit the creation of observers within it at some stage. Both principles are, to a small degree, anti-Copernican, but have been used by Carter to conclude that the Eddington coincidences are support for the usual theory of general relativity, rather than evidence in favour of more exotic theories.

The viewpoint that no exotic theories are needed to explain some of the dimensionless numbers would seem to be valid. The first coincidence explicable in this way is that between the mean number of baryons in a star ($1 \, M_\odot \approx 10^{57}$, or 'about' 10^{60}, protons) and the 3/2 power of the inverse of the number (3c) as understood in terms of the dimensionless gravitational coupling constant. This coincidence has been commented on by Jordan and Bondi (see Carter 1974, p292), and can be obtained as a rough equality between the two numbers on the hypothesis that stars are supported by non-relativistic gas pressure. Outside this regime, stars, if they do not collapse, would be unstable against fragmentation or continuous mass loss. The existence of stability is therefore a prerequisite for the coincidence to hold, but it does not seem to me that this coincidence is particularly striking anyway.

The second large-numbers coincidence considered by Carter is that treated by Dicke, namely the coincidence of the age of the Universe (in suitable units) with the parameters of stars. It has been given essentially the same explanation. A typical star has a luminosity produced by hydrogen burning and is constrained in opacity by Thompson scattering, the connection between the two resulting in the hydrogen-burning lifetime of a main sequence star like the Sun being roughly H^{-1}, or the age of the local Universe. This reasoning is an example of the weak anthropomorphic principle.

The third coincidence is an example of the application of the strong anthropomorphic principle, which aims to explain the rough equality $\rho_U \approx 3H^2/8\pi G$, where ρ_U is the mean mass density of the Universe. This problem is considered at length in §4.5, but Carter has noticed that this relation, plus the second coincidence considered above, gives Eddington's famous coincidence between the number of particles in the Universe (4) and the square of the inverse of the gravitational coupling constant (3c). Carter's reasoning involves the parameters $\gamma_1 \equiv \langle n_b \rangle/T_{bb}^3$ and $\gamma_2 \equiv {}^3K/T_{bb}^2$, where $\langle n_b \rangle$ is the root mean square baryon number, T_{bb} is the temperature of the black-body radiation, and ${}^3K$ is the scalar curvature of the homogeneous space sections of the Universe. If the Universe is not radiation-dominated, there exists a relation between H^{-1}, γ_1, γ_2 and the familiar constants, which gives an upper limit to γ_2 based on the assumption that the Universe is closed. A lower limit to γ_2 can be obtained by adopting the viewpoint (treated in §4.5) that observers can only exist if condensations also exist. This demands that matter has by now decoupled from the radiation, so that T_{bb} is below the Rydberg ionisation energy by a comfortably large amount. The further development from the incipient condensations thus assumed, up to the scale of the observed galaxies, also requires that ${}^3K$ has been non-negative (or else has been very small in magnitude; γ_1 is known to be small). The limits to γ_2 so arrived at roughly restrict the quantities involved to lie in ranges resulting in the third large-numbers coincidence. The coincidence is a result, in short, of the fact that γ_1 and γ_2 are not large. The use of the strong anthropomorphic principle in this way does not rule out the possibility of constructing a theory giving rigid predictions of the numbers involved (for example, Sciama's Machian framework type considered in Chapter 3, in which $\gamma_2 = 0$). It does show that conventional general relativity is correct up to a certain degree, in that it predicts an acceptable model Universe in which the third large-numbers coincidence would be consistent with the existence of observers of a type necessary to observe the relationship.

The three noted coincidences have explanations involving no anthropomorphic argument, the use of the weak anthropomorphic principle, and the use of the strong anthropomorphic principle, respectively. While the employment of the weak principle seems acceptable, I think it is wise to express here a doubt concerning the use or meaningfulness of the strong principle. It does not seem to me that the strong principle qualifies as an explanation of things that carries any degree of conviction. While the weak principle serves as a useful indication of the direction in which existing theory can most profitably be extended, it should be pointed out that any inclination to promote the strong principle to the status of a theoretical explanation of some natural phenomena would be a procedural mistake. It would be all too easy for the strong principle to become an excuse for curtailing the investigation and construction of unconventional gravitational theories. The strong anthropomorphic principle could only be justified as a meaningful hypothesis if there exist, or have existed, other universes where the existence of observers is precluded. One such possibility, discussed by Carter and mentioned elsewhere in connection with entropy-producing processes, is that of the world ensemble, in which there are numerous universes possessing different values of the fundamental constants. In particular, our Universe is characterised by a strong-interaction coupling constant that is only just large enough to bind nucleons into nuclei and provide elements with atomic numbers larger than hydrogen. There may conceivably be other universes in which it could be weaker and in which, therefore, there would be no heavy elements and presumably no observers.

With the purpose of illustrating the possible irrelevance of the strong anthropomorphic principle, it can be mentioned that there is a hypothesis on the nature of stars, based on an idea of Ambartsumian, and extended by Zeldovitch, Vartanyan, Wesson and others (see, for example, Vartanyan 1966) which employs the mass defect concept of gravitational binding energy (analogous to the nuclear packing fraction). One can show that a sphere of baryons with a power-law density profile (index N)

has a total baryon number that can be predicted, if certain assumptions are made about optimising the mass defect. It has been shown that the number of baryons in a star on this simple model is

$$N_* \lesssim N_{\max} \approx \left(\frac{m_p c}{\hbar}\right)^3 \left(\frac{a\hbar^3}{m_p^4 c^3}\right)^{3/N},$$ (4.9)

where a is a constant ($= [ML^{N-3}]$ dimensionally with $a = c^2/G$ if $N=2$), and $N_{\max}$ is an upper limit which would probably be attained in practice. The number N lies in the range $0 \leqslant N \leqslant 3$, and can be connected with the large-scale structure of the Universe. Its observational value is $1 \cdot 7 \lesssim N \lesssim 2$; while the whole account is analytic only if $N=2$. An appeal to observation and mathematical elegance therefore suggests that $N=2$ should be taken, in which case equation (4.9) gives $N_* \simeq (c\hbar/m_p^2 G)^{3/2}$. The number $(c\hbar/m_p^2 G)$ is, of course, dimensionless, and for normal stars such as the Sun it is observed that $N_\odot \simeq N_* \approx 10^{57}$. The explanation for $N_\odot \simeq (c\hbar/m_p^2 G)^{3/2}$ on this hypothesis (which is general relativistic, but of a type that is not widely known), lies in the packing of baryons close together in a massive sphere with a value of the free parameter (N) inferred from cosmology. The hypothesis makes no appeal to anthropomorphic principles, even though it provides a basis for some of the Eddington coincidences as they are understood by Carter.

The counter-example outlined above should suggest caution in the use of the strong anthropomorphic principle. The weak anthropomorphic principle, alternatively, is more acceptable and serves the important purpose of showing that some of the coincidences implicit in the values of the constants of physics could well be consequences of our own existence as salient observers.

4.5. Isotropic Universes

A philosophy similar to that of Dicke and Carter has been adopted by Collins and Hawking (1973) in an attempt to understand why the relation $GM_U/R_U c^2 \approx 1$ appears to hold (cf §3.2), and why observers are present at this stage in the evolution of the Universe. Their reasoning is as follows.

Human beings exist because small perturbations grew at some stage in the history of the Universe, notably into galaxies, and yet the (presumed cosmological in origin) 3 K microwave background is remarkably isotropic and sets stringent limits on large-scale perturbations in the mass density. The isotropy of the Universe and the existence of observers are therefore closely related phenomena.

Assuming that the Universe is roughly Friedmannian, analyses of the growth of homogeneous, but anisotropic, perturbations can be carried out. Inhomogeneous perturbations are not considered because they would produce small-scale anisotropy in the microwave background which is not observed (Hawking 1974). The model universes that admit condensations (galaxies) at the present epoch, but are isotropic on the large scale, can be singled out by studying the response to perturbations of all the feasible models.

In the $k = +1$ (closed) Friedmann Universe, the perturbations have to be Bianchi type IX (the Bianchi classification is based on symmetry properties; types I, V, VII_h, VII_0 and IX are concerned, where h is a five-dimensional class label). In the $k = 0$ (parabolic) case, the perturbations have to be of Bianchi types I or VII_0, the former representing different rates of expansion in three mutually orthogonal directions in Euclidean space sections, while the latter involves rotation and shear. In the $k = -1$ (hyperbolic) model, the perturbations can be of Bianchi types V or VII_h. The $k = +1$ perturbations decrease in amplitude until the Universe reaches its maximum radius and recollapses. The $k = 0$ perturbations of both types die away progressively. The $k = -1$ perturbations die away if they are of type V, but some of those of type VII_h grow in the later, low-density stages of expansion. The cosmological constant has been taken as $\Lambda \equiv 0$ in all of these cases.

The implication of the foregoing results is that the Universe will only be isotropic on the large scale with condensations on the small scale (galaxies) if its velocity of expansion is near to the critical value characteristic of parabolic models. If the expansion velocity were much larger, anisotropic perturbations would have grown by now; if much slower, the Universe would tend to recollapse, and there would not have been time for the anisotropy to be damped out. If Λ is indeed zero, this implies

100

that the mean density of the Universe (ρ_U) is very near to the critical density with a reasonable value of Hubble's parameter H ($= 50$–100 km s^{-1} Mpc^{-1}), this density being $\rho_U = 3H^2/8\pi G$. This result, from an abstract viewpoint, would mean that the chance of life developing in a given universe would be very small, since Collins and Hawking (1973) have shown that the set of spatially homogeneous cosmological models that tend to isotropy as $t \rightarrow \infty$ is of measure zero in the space of all spatially homogeneous models. Robertson–Walker models possess isotropy that is unstable to homogeneous, anisotropic perturbations; only a small set of initial conditions gives universes which are isotropic to within limits presently set by observations. Bonner (1974) has been able to show that many spherically symmetric, inhomogeneous models tend to homogeneity and isotropy with time (see also Davies 1974a), so that the likelihood of life evolving from conditions that form a set of measure zero may not be as unlikely as might initially have been thought. Nevertheless, the isotropy conditions do leave a strong impression that life as it has evolved from a set of measure zero must be an uncomfortably rare occurrence. One way out of this dilemma is to suppose that an infinite number of universes with all possible different conditions actually exist, but that life only develops in those expanding just fast enough to avoid recollapse, these models also tending to isotropy. Of course, anisotropies in the early Universe might have been reduced by neutrino viscosity, assuming that the initial irregularities were not too large (Misner 1968), but only in models which have exactly the escape velocity can galaxies grow easily.

The many-universe picture has been considered by Dicke and Carter as a viable explanation of results of the sort being discussed. The consideration of our world as a single member of an ensemble is familiar from the Everett interpretation of quantum mechanics, in which all quantum alternatives are realised simultaneously. The acceptability of the Everett scheme rests on the fact that the overwhelming majority of such worlds have entropy-increasing regimes; only those satisfying this criterion, and being compatible with life, are observed. The Everett interpretation is discussed by Davies (1974c, pp172, 199), who shows that the time asymmetry observed in the actual local world is probably a consequence of the continuous formation of branch systems. It is likely, however, that further insight into the meaning of entropy and the Everett scheme will be gained by using results of Hawking, Beckenstein and others on black-hole thermodynamics (see, for example, Hawking 1975). Black holes as inferred from classical general relativity have an entropy expressed by $S_{bh} = \pi k c A_{bh}/2hG$ and an equivalent temperature $T_{bh} - c/8\pi G k M_{bh}$, where A_{bh} and M_{bh} are the surface area and mass, respectively, and k is Boltzmann's constant. When quantum effects are taken into account, it is predicted that black holes smaller than 10^{-13} cm should have vanished by radiative mass loss. Black holes become hotter as they radiate, and this corresponds to a negative specific heat that has important consequences for the meaning of entropy and the large-scale thermodynamic evolution of the Universe. In the cosmological context, the isotropy of the Universe and the existence of life may both be due to the expansion of the Universe at about the Einstein–de Sitter critical rate. We would not be here to observe the Universe if it were not like this.

4.6. Rotating Universes

The Dirac relationship expressed in the form $m_e c^3/He^2 = e^2/Gm_e^2$ has been used by Cavallo (1973) in the more exact form $m_e c^3/He^2 = \xi_C e^2/G\eta_C m_e^2$, where ξ_C is a constant, and the radius of the Universe is $R_U \equiv \eta_C c/H$, so that Hubble's parameter is $H = \eta_C G m_e^3 c^3/\xi_C e^4$. Guided by an empirical relationship between the angular momenta and volumes of bodies from the sizes of atoms up to galaxies ($L \propto V^p$, $p \simeq 1$), Cavallo calculates that if the Universe is rotating, its angular momentum is $L_U = \hbar m_e^3 c^9 \eta_C^3/2H^3 e^6$. A connection between L_U and H can be formed by employing the equation

$$\frac{H^2}{2} + \frac{\Omega_U^2}{3} - \frac{\Lambda c^2}{6} = \frac{4\pi G\rho_U}{3} + \theta_U \qquad (4.10)$$

of Heckman and Schücking (1956), which describes a rigidly rotating, expanding Newtonian Universe (where Ω_U is the angular velocity, Λ is the cosmological constant, ρ_U is the mean density, and θ_U is the energy constant). Assuming that the

Universe is a sphere, its angular momentum is approximately

$$L_U = \frac{2M_U R_U^2 \Omega_U}{5} = \frac{2\chi_C \zeta_C \eta_C^5 c^5}{5GH^2}, \qquad H = \frac{5G\hbar m_e^3 c^4}{4\chi_C \zeta_C \eta_C^2 e^6},$$

(4.11)

where χ_C is another constant introduced by Cavallo via $\rho_U = 3\chi_C H^2/4\pi G$ and $\zeta_C \equiv (1 - \eta_C^2)^{1/2}/\eta_C$. Employing equations (4.10) and (4.11), it is possible to eliminate H and find α ($\equiv e^2/\hbar c) = 5\xi_C/4\chi_C \eta_C^3 \zeta_C$, which can be a number of the observed size of the fine-structure constant.

In the sense of attempts to understand the interrelationship of the Eddington numbers, the interrelationships promulgated by Cavallo are rather complex, lacking the intuitive appeal of the simple Dirac equation. The hypothesis that the Universe is rotating is very suspect in that strict limits on the rotation have been set by observations of the microwave background (Collins and Hawking 1973) and the x-ray background (Wolfe 1970), although the latter limit depends on certain assumptions regarding the density of intergalactic hydrogen, and might be relaxed. The microwave background is, of course, expected to be anisotropic at some level, due to radiation from fine-structure transitions in common ions within the Galaxy and in other galaxies for wavelengths less than 0·5 mm (Petrosian et al 1969) and to processes involving microwave emission from dust at longer wavelengths (Wesson and Lermann 1976, Duley 1976, Clegg et al 1976). However, the isotropy of the background at reasonable intensities nevertheless seriously restricts any large-scale rotation and renders doubtful any explanation of the Eddington numbers on this basis.

4.7. Elementary Particles

Several explanations of the Eddington relationships between the fundamental constants have been proposed on the bases of elementary particle physics and statistical mechanics. Unlike the cosmological explanations, some of the reasoning underlying the proposed explanations is not complete, although those accounts treated below, which make explicit use of parameters connected with the age and rate of expansion of the Universe, are fairly well developed. This section examines six hypotheses involving elementary particle theory and cosmology.

(1) The condition that the Universe was homogeneous in its early stages has been used by Clutton-Brock (1974a) to explain Dirac's coincidence as an elementary particle effect. The Universe could have been homogeneous because of causal processes which operated when the Compton wavelength of the pion (λ_π) was larger than the particle horizon (Clutton-Brock 1974b). Such a hypothesis would predict the condensation of superclusters at a redshift $z \simeq 11$, followed by fragmentation into clusters, galaxies, etc. At that epoch, the repulsion between baryons would have extended beyond the particle horizon, violating classical causality, but allowing the growth of density perturbations into astrophysical systems, while preserving the isotropy of the black-body background. The existence of observers at the present time leads to the assumption that condensations did grow at $z \approx 10$, with the perturbations increasing from an isotropic epoch $t_I = \lambda_\pi/c$ when the number of baryons within the horizon was of the order of 10^{80}. The particle physics of the growth of the perturbations is closely connected with the numerical properties that $e^2/\hbar c \simeq 1/137 \simeq m_e/m_\pi$ and $(T m_e c^3/e^2) \approx 10^{40}$, so that the coincidence commented on by Dirac can be connected with the growth of primaeval perturbations, even if of a somewhat unlikely type. The appearance of two of the Eddington numbers in perturbation analysis will be seen below to occur also in the Hagedorn thermodynamic picture, although the reason is obscure. (Harrison (1972) has examined particle physics in the very early Universe.) It would seem that numbers of the order of 10^{40} and 10^{80} are a general consequence of a range of possible processes in the early Universe, even if their precise relationship to the other numbers and present-epoch data cannot be clearly perceived.

(2) Particle physics in the cosmological sense has been used by Narlikar (1974c) to explain the same relationship treated by Clutton-Brock. Narlikar's hypothesis is based on the idea that a fluctuating electrostatic field of universal extent can create positrons and electrons, and the whole system remains electrically neutral. The variation of the world line of a lepton of

mass m_1 and charge e gives the relation

$$m_1 u_i = -eA_i \qquad (4.12)$$

at the beginning or end of a world line (where A_i is the four-potential of the rest of the particles in the Universe, and u_i is the four-velocity of the lepton). If electrically charged particles in the Universe only balance each other statistically, there will be fluctuations in number of size $\sqrt{n_U}$, where n_U is the number of charges in a region of the Universe of size R_U, producing an overall electrostatic potential of the order

$$A_4 \approx \pm \frac{\sqrt{n_U} e}{R_U}, \qquad (4.13)$$

where $n_U \approx 10^{80}$ in practice. Combining equations (4.12) and (4.13) for a lepton created at rest in the substratum,

$$m_1 c^2 = \left| \frac{\sqrt{n_U} e^2}{R_U} \right|, \qquad (4.14)$$

which might imply that leptons of different masses are created as $H(\equiv \dot{R}_U/R_U)$ varies. The last equation is, in fact, one of the Eddington coincidences, since it can be re-written as

$$\frac{R_U}{(e^2/m_1 c^2)} \simeq \sqrt{n_U}, \qquad (4.15)$$

stating that the ratio of the radii of the Universe and an electron ($m_1 \equiv m_e$) is a number of the order of 10^{40}.

The second relationship of this type used by Dirac can be obtained in a more standard manner via the adoption of equation (4.8) in the form $M_U = 4\pi R_U^3 \rho_U / 3 \simeq c^3/HG$, so that $M_U/m_p \simeq n_U$, with $n_U m_p \approx c^3/HG \simeq c^2 R_U/G$. Using the last relation to eliminate R_U from equation (4.15) gives the second Dirac coincidence as

$$\frac{e^2}{G m_p m_e} \simeq \sqrt{n_U} \approx 10^{40}.$$

The smallness of the gravitational force in an atom compared with the electrostatic force is therefore a consequence of the smallness of an electron compared with the Universe (equation (4.15)) and the existence of many electrons, combined with the assumption that the Universe is massive and attains the critical density. The connection of the Eddington numbers with lepton electrodynamics is discussed further by Narlikar.

(3) The full implication of lepton creation as it concerns the other Eddington numbers has not been developed, but Teller (1972) has remarked that, if the negative gravitational energy and positive rest mass energy of particles are added, the sum is approximately zero. New particles can therefore be created without violation of the conservation of energy. Both Teller (1972) and Wataghin (1972) were concerned with the relation between particle processes and the early stages of the big bang, as first advocated by Gamow. (The latter considers that the transfer of particle energy to gravitational energy is responsible for the expansion of the Universe, with the redshift resulting from the work done by particles of zero rest mass against the gravitational field of the Universe.) The interactions of particles in the big bang to produce elements by nucleosynthesis was investigated by Gamow (1948), who found that the mass of condensations, M_c, fragmenting out of the primaeval medium was composed of a series of fundamental constants, the actual value being

$$M_c \simeq \frac{2^{31/8}\, 5^{7/4}\, \pi^{5/4}\, e\hbar^{5/4}\, \epsilon_D^{5/4}}{3^{17/8}\, m^{15/4}\, c^{5/4}\, G^{7/4}} \qquad (4.16)$$

and of the order $3 \times 10^7\, M_\odot$, where ϵ_D is the binding energy of a deuteron (2·2 MeV). Alpher and Gamow (1968) and Alpher and Herman (1972) found that some of the Eddington coincidences could be explained without time dependence of the parameters, basically because the ratio of the heat capacities of radiation and matter depends on the fundamental constants in a conveniently factorisable manner. This ratio, which does not alter during adiabatic expansion of the Universe, is only a function of the cross-over time, which is defined as the epoch when the densities of matter and radiation were equal; that is, $t_c \simeq 10^6$ yr $= 3 \times 10^{36}$ tempons, where 1 tempon $\equiv e^2/m_e c^3 = 0.94 \times 10^{-23}$ s, compared with $e^2/Gm_p^2 \simeq 1 \times 10^{36}$, so that the heat capacity ratio is only a function of t_c or of e^2/Gm_p^2. When numerical values are substituted with the known data on the microwave background, there emerges the figure 10^9 for the photon/baryon number ratio. This suggests that the ratio as

106

now observed depends on the familiar laws of thermodynamics applied in the early Universe, with the values of the constants substituted as they are known at the *present* epoch. Rees (1972) has reviewed the status of this last number as it relates to the Eddington numbers.

(4) Elementary particle physics and statistical mechanics have been employed by Lawrence and Szamosi (1974) in formulating an explanation of the Eddington coincidences. They observe that, if there is a population of n_U particles in the Universe, statistical mechanics predicts that there should be a spread in mass around the equilibrium value of the amount

$$\frac{\delta m}{m} \approx \frac{1}{\sqrt{n_U}}, \qquad (4.17)$$

where m is the mean particle mass. Further, if T is the age of the Universe, quantum mechanics shows that it is possible to detect a mass difference $\hbar/c^2 T$ over a time T, so that δm must be less than this critical value and one might expect

$$\delta m \simeq \frac{\hbar}{K_p c^2 T}. \qquad (4.18)$$

Here, K_p ($\gtrsim 1$) is a dimensionless constant whose size is connected via the mass difference δm with the Pauli Principle in an anthropomorphic way, since the existence of human beings depends on the existence of nuclei and so on the near equality of the proton and neutron masses. To understand the background to the hypothesis of Lawrence and Szamosi, one should appreciate that our existence as observers depends on the long-term variability of the Pauli Principle.

The lifetime of the Exclusion Principle has been studied by Reines and Sobel (1974) who searched for the disappearance of K-shell electrons in iodine atoms. The fraction of iodine atoms that have altered due to the breakdown of the Pauli Principle since the origin of the Solar System is less than 10^{-9}, and a lifetime against violation of the Principle can therefore be set that is at least equal to the age of the Universe and *may* be very much longer ($\gtrsim 2 \times 10^{27}$ s $\approx 10^{20}$ yr). The half-life of the proton against decay into energetic particles is greater than 2×10^{30} yr (Reines and Crouch 1974), and decays in ^{39}K nuclei could probably be used to study non-fragmentary decay on time scales up to 2×10^{23} yr (Rosen 1975). The half-life may not, however, be infinite since gauge theories such as that of Pati and Salam, which unify the weak, electromagnetic and strong interactions (Rosen 1975, p774), predict a proton half-life in the range 10^{28}–10^{34} yr. However, Veltman has supported criticism of such theories by Linde (Veltman 1975), since they predict unacceptable cosmological forces of strengths 10^7 times those of gravity (see §2.8). These originate from terms that are negligible on the nuclear scale, but large on the astronomical scale, and enter because of the negative energy of the vacuum and the existence of Higgs mesons (which are the result of giving masses to particles in the gauge theories that would otherwise be massless). Higgs (1975) has sought to counter this argument by appealing to the existence of super-symmetric gauge theories, in which the vacuum is unchanged, and to the undefined nature of the zero level of energy in quantum field theory. The possibility of a finite lifetime for the proton and other elementary particles is therefore undecided.

Lawrence and Szamosi, after noting that the existence of nuclei from at least the beginning of the Solar System (age $\approx T$) depends on the nuclear indistinguishability of protons and neutrons, or equivalently on the validity of the Pauli Exclusion Principle, make the explicit assumption that the Universe is just gravitationally bound, as in equation (4.8). Thus, if m is the mass of an elementary particle, of which there are n_U in a Universe of radius R_U,

$$R_U \approx cT \approx \frac{GM}{c^2} \approx \frac{Gmn_U}{c^2}. \qquad (4.19)$$

This assumption (§3.2) is compatible with most hypotheses concerning Mach's Principle. Combining equations (4.17), (4.18) and (4.19) gives the mass of an elementary particle in terms of fundamental constants:

$$m \approx \left(\frac{\hbar^2}{K_p GcT} \right)^{1/3}. \qquad (4.20)$$

With $T \simeq 10^{10}$ yr, this gives $m \approx K_p^{-1/3} \times 10^{-25}$ g, which fixes an upper limit to the mass of an elementary particle (to be consistent with the existence of human life in the Universe) of

$m \lesssim 10^{-25}$ g, since $K_p \gtrsim 1$. Larger values of K_p would give masses lying in the observed range of 10^{-27}–10^{-24} g.

Lawrence and Szamosi assume (Reines and Sobel 1974) that the lifetime of the Pauli Principle for inner-shell electrons may, in fact, only be $T \approx 10^{10}$ yr, corresponding to $K_p \approx 10^5$, and giving from equation (4.20) the particle mass as $m \simeq 10^{-27}$ g $\simeq m_e$. The relation in equation (4.20) was also noticed by Weinberg (1972, pp619–20) who regarded it as a coincidence that could have implications for microphysics and cosmology. The other Eddington numbers emerge from the Lawrence and Szamosi analysis by eliminating G from equations (4.19) and (4.20), giving $R_U \approx \sqrt{n_U}$ ($\hbar/mc$). Eliminating R_U gives $m \approx (\hbar c/G)^{1/2} n_U^{-1/4}$, where $(\hbar c/G)^{1/2}$ is the Planck mass. The fine-structure constant α appears implicitly since

$$\frac{\hbar c}{Gm^2} = \left(\frac{e^2}{Gm^2}\right)\frac{1}{\alpha} \approx \sqrt{n_U}. \tag{4.21}$$

The last equation, together with the rough equalities $R_U \approx \sqrt{n_U}$ ($\hbar/mc$) and $m \approx (\hbar c/G)^{1/2} n_U^{-1/4}$, are usually referred to as the cosmological coincidences. The fine-structure constant can be estimated to an order of magnitude by considering the electromagnetic self-energy of an electron, with the ultraviolet divergence removed by the inclusion of the gravitational self-energy following Salam, this being

$$E_{self} = \left(\frac{me^2c}{\hbar}\right) \ln n_U. \tag{4.22}$$

If the variation in E_{self} by the spread in mass δm is undetectable, the fine-structure constant should be

$$\alpha \equiv \frac{e^2}{\hbar c} \approx \left(\frac{K_p}{K_p'}\right)\frac{1}{\ln n_U} \tag{4.23}$$

where K_p' is another dimensionless constant. Provided $K_p \approx K_p'$, this gives the order of magnitude of α as $\approx 1/\ln(10^{80})$, approximately as observed.

The hypothesis of Lawrence and Szamosi, while attractive, makes certain assumptions that are really in need of further justification. (i) Equation (4.18) as it stands is a statement of the Heisenberg Uncertainty Principle, and the senses of minimum and maximum observable differences, δm, as required by

109

the Pauli and Heisenberg Principles, would seem not to be in agreement with regard to the limiting size of K_p. (ii) The closed-Universe assumption should be relaxed, if feasible, to examine the sensitivity of the hypothesis to the admissibility of cosmological evidence in favour of an open Universe. (iii) The assumption of a finite lifetime for the breakdown of the Pauli Principle for inner-shell electrons does not seem pertinent or, indeed, necessary. The explanation of the Eddington numbers on the lines discussed by Lawrence and Szamosi could be improved if these points were taken into consideration.

(5) A hypothesis involving elementary particles has been proposed by Landsberg and Bishop (1975a) as an alternative to that involving a statistical spread of elementary particle masses. Landsberg and Bishop assume that (i) only the parameters $\hbar$, c, G and H appear in the problem; (ii) the value of H/G is independent of time, and (iii) one or more particles exist, whose bare rest mass, m, is independent of time. Assumption (i) implies that any mass will occur in the form

$$m \propto \hbar^p H^q G^r c^s,$$

giving

$$m(b) = k(b) \left(\frac{\hbar^3 H}{G^2}\right)^{1/5} \left(\frac{c^5}{\hbar H^2 G}\right)^{b/15}, \qquad (4.24)$$

where p, q, r, s are constants and b is an unidentified parameter that fixes elementary particle masses via dimensionless constants $k(b)$. The parameter b lies in the range $-6 \lesssim b \lesssim 9$, and the lower limit is fixed by the least mass difference measurable in the period H^{-1} since the big bang; the upper limit is defined by less-clear cosmological considerations. The Eddington number of the order of 10^{80} is calculated as

$$\frac{m(9)}{m(-1)} \approx \left(\frac{c^5}{GH^2\hbar}\right)^{2/3} = 24 \times 10^{80}. \qquad (4.25)$$

The time-independent particle mass is $m = m(-1) \approx 10^{-25}$ g, and is therefore equal to 0·4 of the rest mass of the charged pion. A choice of $b = 3/2$, in conjunction with this basic particle mass, gives the other Eddington number

$$\left[\frac{m(3/2)}{m(-1)}\right]^2 \approx \frac{\hbar c}{Gm^2} \approx 10^{40}. \qquad (4.26)$$

The calculation of dimensionless numbers in this hypothesis would seem to be somewhat arbitrary, in so far as one might choose any suitable value of the parameter b. It also depends on the assumption (i) concerning G. This is in accordance with Dirac, who implicitly took $G(t)=(\text{constant})H(t)$, but is more restrictive than G dependencies such as those occurring in the Hoyle–Narlikar theory. Nevertheless, the cosmological hypothesis with which Landsberg and Bishop complement their elementary particle hypothesis is less open to criticism. A time dependency of G is introduced into Newtonian cosmological models by using a principle of impotence (Landsberg and Bishop 1975b), which states that it is impossible to obtain the radius of the Universe by a study of local dynamics. The energy equation in Newtonian cosmology can be written as a relation involving the scale factor $S(t)$, the mean density $\rho_U(t)$, a cosmological constant Λ, and an integral over a local domain of space that includes $G=G(t)$. Conservation of matter allows one to eliminate $\rho_U \propto S^{-3}$ in comoving coordinates, since $H=\dot{S}/S= -3\dot{\rho}_U/\rho_U$, and the resulting relationship expressing conservation of energy can be put in the form of a differential equation involving $S(t)$ and $\dot{S}(t)$. Imposing the principle of impotence obliges $S(t)$ to be unobtainable locally, so that the law of gravity takes the usual form only if $G(t) \propto [S(t)]^{-j}$, where j is a constant. Data on the value of Hubble's parameter H, the deceleration parameter and the age of the Universe constrain j to lie in the range $-1 \lesssim j \lesssim 4$. A choice of j fixes G explicitly, since

$$G(t)=\exp\left[-j\int^t H(t)\, dt\right], \qquad (4.27)$$

and the gravitational interaction energy is of the form

$$E_g(r,\, t)= -m_1 m_2[(1/r)+Dr^j]G(t), \qquad (4.28)$$

where D is a constant. This agrees with the Newtonian law in the first term. The second term can be chosen to alleviate the 'missing-mass' problem in astrophysics, for example, since

$\rho(\text{here})/\rho(\text{Friedmann}) = 1/(j+1)$. The Dirac cosmology corresponds to $\dot{G}/G = -3H$, with a somewhat compromising 5×10^9 yr as the age of the Universe. The finite $\dot{G}/G$ value of Van Flandern, discussed in §5.8, is equivalent to $j = 1\cdot4 \pm 0\cdot9$.

The connection between these various cosmological considerations and the impossibility of locally evaluating $S(t)$ would seem to be important, since it has always seemed to me that $S(t)$ as such, being ill-defined and of doubtful conceptual meaning even in the best-understood cosmologies, should not explicitly appear in local astrophysical theory except via certain more clearly defined functions of itself. In Robertson–Walker cosmology, for example, Friedmann's equations show that $S(t)$ only appears as meaningful functions of itself if the curvature constant is zero, and this may provide a link between the account of §4.5 and the work of Landsberg and Bishop.

(6) The last hypothesis to be examined in this section concerns Hagedorn's statistical thermodynamics of strong interactions among particles in the big bang. One of the main objectives of this theory is to determine $\rho_H(m)$, the number of elementary particles in the range of mass $m \to m + dm$ (or the mass spectrum). A self-consistent argument can be used to obtain $\rho_H(m)$, as shown by Hagedorn (1974, p273). The steps are as follows. (i) Suppose that a strong-interaction thermodynamics exists, meaning in the cosmological sense that fireballs exist or have existed. (ii) These can have any mass, and once $\rho_H(m) \, dm$, the number of different species of hadron in the range $m \to m + dm$, is known, one can calculate the level density $\sigma_H(M) \, dM$, which is the total number of states of the *whole* composed object with mass in the range $M \to M + dM$. (iii) The continuum of fireball masses shows that fireballs become resonances at low masses, so resonances $\equiv$ fireballs. (iv) A description including all resonances in $\rho_H(m)$ must therefore include fireballs themselves (as fireballs contain resonances). (v) The mass spectrum $\rho(m)$ must therefore include not only resonances, but also $\sigma(M)$, the two becoming the same function at large masses and so giving a limit within the thermodynamical framework resulting in (vi) a unique asymptotic solution, namely,

$$\rho_H(m) \sim (\text{constant}) \, m^{-3} \exp{(m/T_H)}. \qquad (4.29)$$

(vii) The whole $\rho_H(m)$ is now known, since its low-mass behaviour has just been fixed. (viii) If $\rho_H(m)$ is known, the thermodynamics is completely determined, closing the circle.

The Hagedorn theory has been much used in cosmology. It predicts a universal maximum temperature of $T_H \simeq 1\cdot6 \times 10^{12}$ K ($\equiv 140$ MeV$\equiv 1$ pion mass). It agrees with the observed spectrum of about 1000 known particles (counting, individually, factors due to differing spin, isospin, and particle/antiparticle nature), most of which are resonances. The theory has been employed by Kundt, and by Carlitz *et al* (see Hagedorn 1974, p282) in an account of galaxy formation in the primaeval fireball. The theory of the latter workers is a zero baryon number model (equal amounts of matter and antimatter), and can explain the formation of galaxies as condensational growths from the smoothest possible state at 10^{-4} s after the creation event. The theory can probably be reworked to explain the origin of galaxies in a predominantly matter universe. In the zero baryon number model, fluctuation calculations unexpectedly produce the two numbers 10^{38} and 10^{76}, which are close to the Eddington numbers. The first represents the ratio of the size of a fluctuation cell (of a volume such that the average energy of the fluctuation is unity) to the volume of a pion in the fireball. The second represents the ratio of the equivalent masses. Both are evaluated at 10^{-23} s, the beginning of the hadronic era, when the horizon is just equal to the size of an elementary particle of pion mass. If V_F is the fluctuation cell volume, and M_F the corresponding mass,

$$V_F(10^{-23} \text{ s}) = 10^{38} \, v_\pi;$$
$$M_F(10^{-23} \text{ s}) = 10^{76} \, m_\pi. \tag{4.30}$$

The physical reason for the appearance of these numbers is not easy to discern, but it must be closely connected with the Hagedorn thermodynamics.

4.8. The Fine-Structure Constant and G

Attempts to elucidate the meaning of the Eddington numbers have often been linked with a wish to explain the size of that famous dimensionless number, the fine-structure constant. Limits on the variation of the latter (§4.9) can be used to constrain variations in G once a relationship is found between the two parameters.

The possibility of connecting atomic theory (involving $\alpha \equiv e^2/\hbar c$) and the gravitational constant (which is usually taken as diagnostic of the treatment of some large-scale problem) was first formulated by Landau (1955). Landau's reasoning starts from the existence of an unrenormalised fine-structure constant, α_0, whose value is such that $\alpha_0 \ll 1$. The renormalised (that is, familiar) fine-structure constant (α) is then given by

$$\alpha = \frac{\alpha_0}{1 + (\alpha_0 \nu_L / 3\pi) \ln(m_c{}^2/m_1{}^2)}, \qquad (4.31)$$

where m_1 is the mass of an elementary particle, ν_L a number depending on the spin $(0, \frac{1}{2})$ of the differently charged elementary particles, and m_c is a cut-off mass. Arnowitt et al (1960) have shown that, when the gravitational effect of the classical Coulomb self-energy of a point charge is taken into account, the total rest mass is finite and depends only on the charge, not on the bare mechanical mass (so that a particle acquires mass only when it has non-gravitational interactions with fields of nonzero charge). The gravitational force holding the charge distribution of an elementary particle together is thus counteracting the electrical forces in the system, and the charge of an elementary particle (Peebles and Dicke 1962) is confined to a region of size

$$\frac{eG^{1/2}}{c^2} \approx 10^{-34} \text{ cm.} \qquad (4.32)$$

This can be used to derive m_c, the cut-off mass, for use in equation (4.31), and one obtains

$$\alpha^{-1} \simeq (\nu_L / 3\pi) \ln\left(\frac{\hbar c}{Gm_1{}^2}\right). \qquad (4.33)$$

Here, $Gm_1{}^2/\hbar c$ is a dimensionless number of the order of 10^{-40} when a suitable particle mass is substituted for m_1. The relation in equation (4.33) provides a hypothetical connection between α and G. Variations in G then imply variations in α

and vice versa, and the relative sizes are given by

$$\frac{\delta\alpha}{\alpha} \simeq 10^{-2}\,\frac{\delta G}{G}. \tag{4.34}$$

This form of procedure has been used by Peebles and Dicke (1962) to calculate that since α changes by an amount smaller than three parts in 10^{13} yr^{-1}, then G changes by less than two parts in 10^{11} yr^{-1}. The latter figure is five times smaller than the value expected from the original hypothesis of Dirac concerning the time variation of G, but is just about equal to the value expected from the Brans–Dicke theory. Limits on the time variability of α (§4.9) are not in disagreement with the predicted time variability of G. Dicke (1953, 1960a, b), in discussing the connection between variations in α and G, finds that predictions about the space variation of G expected from Machian theories (§3.4) are not contradicted by experiments on the variability of α. It is probable that close limits on the variability of G from other branches of physics would be best used to restrict the variability of the fine-structure constant.

4.9. Dimensional Constants and Coupling Constants

Changes in dimensional constants other than G have not been extensively sought for, with the exception of attempts to detect a time dependence of h, Planck's constant.

Harwit (1971) has given an extensive account of possible methods of determining a time or space dependence of h. The absence of intergalactic absorption of light emitted from distant sources such as QSOs (see Bahcall *et al* 1967 for an account of absorption lines in one distant source), might be due to there having been an ancient value of Planck's constant $(h + \delta h)$ which was different from the present one (h). Observations of 3C 273B indicate that $\delta h/h < 10^{-5}$ for this source. (A common practice in quantum field theory is to choose units such that $h \equiv 1$, just as one can in principle put $G \equiv 1$, but for obvious reasons this practice can cause problems and is not followed here.) Harwit has also discussed a test, previously proposed by Bahcall and Salpeter (1965), in which light from a distant source is measured in dispersion by using a prism and a grating. The prism measures the energy, ϵ, of a photon, while the grating measures the wavelength, λ. The product is

$$\epsilon\lambda = (h\nu)(c/\nu) = hc, \qquad (4.35)$$

and by investigating whether $\epsilon\lambda$ is different for sources at different distances, the variability of h might be tested. It was shown by Harwit, however, that due to the way in which photons propagate in homogeneous, isotropic cosmological models, a pure time dependence of h cannot be directly detected in this way if general relativity holds, although equation (4.35) might be used in other contexts.

Harwit further considered the limits on the variation of h set by Wilkinson, who investigated α decay rates. Wilkinson's observation was that radioactive decay ages agree roughly with the inferred age of the Galaxy, so that the α decay constant cannot have changed by more than a factor of three or four in the last 10^{10} yr or so. This implies, in turn, that e, h and c are probably not altering faster than one part in 10^{12} yr^{-1}, with the meson coupling constant changing comparatively slowly and the β coupling constant changing by less than one part in 10^9 yr^{-1} (Wilkinson 1958). This is a sensitive test, because the decay constant is exponentially dependent on atomic parameters:

$$\text{decay constant} \propto \exp\left(-\frac{8\pi^2 Z e^2}{h\nu}\right), \qquad (4.36)$$

where Z is the atomic number of the nucleus being considered, e is the electron charge and ν is the velocity of the α particle at emission. Wilkinson gave arguments for considering the quantities other than h in the exponential to be non-varying, since this combination of atomic parameters is dimensionless. (It should be noted, however, that the 'conspiracy hypothesis' discussed in §4.10 would partially circumvent Wilkinson's arguments, if it were true.) Wilkinson thus concluded that h is not varying by more than one part in 10^{12} yr^{-1}. As Harwit (1971) has noted, however, h is, in one sense, expected to vary in space, if not in time, by a small amount if general relativity is valid, since Infeld (1963) has shown that h is a function of the metric properties of space in general relativistic universes.

Infeld has drawn an analogy between light travelling in a gravitationally static space and light rays propagating in an

116

inhomogeneous medium. If v is the frequency and μ the inertial energy of a photon, its motion is equivalent to that of a moving particle with zero rest mass, so that

$$g_{ij}\dot{x}^i\dot{x}^j = c^{-2}(ds/dt)^2 = 0, \qquad (4.37)$$

where (·) denotes $c^{-1} \, d/dt$ and $x^4 = ct$. The equation of a geodesic for the light ray gives

$$v \propto \mu\dot{x}_4 = \mu\dot{x}^4 g_{44} = \mu g_{44} = \mu c'^2, \qquad (4.38)$$

where c' is a kind of local velocity of light in units of an asymptotic value c. (The frequency v of a photon is a constant along a geodesic, but the wavelength λ is not, since $\lambda \propto c'/v$.) Denoting a suitable constant by h_0, the last equation gives

$$\mu\dot{x}_4 = \mu c'^2 \equiv h_0 v; \qquad \mu = h_0 v c'^{-2} = h v$$

so that

$$h = h_0 c'^{-2}. \qquad (4.39)$$

The quantity c' is $c' = \sqrt{g_{44}}$, and is a function of space–time, so that in a sense, Planck's constant depends on the local geometry. Harwit (1971) has suggested that one should look at pulsars to see if h varies by an anomalous amount in strong gravitational fields.

By using cosmological methods, it is possible to look for a variation in h both in time and space, since the measured red shift z of sources comprises a combination measure of remoteness in space and in time. Noerdlinger (1973) has compared the Rayleigh–Jeans portion of the $T = 2.7$ K microwave background with the turnover in the complete spectrum. The first gives kT, with no knowledge of h involved, while the turnover gives kT in combination with h (k is Boltzmann's constant). Since the two data agree, it can be said that h has not changed by more than $\delta h/h \simeq 0.3$, corresponding to a possible 30% error in the temperature $T \simeq 2.7$ K over the region $1000 > z > 0$. ($z = 1000$ represents the last scattering surface, near the time origin of the Universe.) On the basis of a Robertson–Walker model, Noerdlinger derived the limit

$$\frac{d[\ln{(hc)}]}{dz} \lesssim 3 \times 10^{-4}. \qquad (4.40)$$

Bahcall and Salpeter (1965) had earlier found the limit

$$\frac{d[\ln{(hc)}]}{dt} \lesssim 10^{-10} \text{ yr}^{-1} \tag{4.41}$$

by data on the non-discrepancy at two different wavelengths of atmospheric absorption measures of the energy of the photons from a distant source. The limit set by Bahcall and Salpeter corresponds to

$$\frac{d[\ln{(hc)}]}{dz} \lesssim 1 \tag{4.42}$$

which is considerably less accurate than the result (equation (4.40)) of Noerdlinger. It is of interest because the two tests are conceptually different. Bahcall and Salpeter also examined multiplet splittings in QSOs to obtain a limit on the fine-structure constant, α, of

$$\frac{d[\ln{(\alpha^2)}]}{dt} \lesssim 10^{-11} \text{ yr}^{-1}. \tag{4.43}$$

The limits on the z and t dependence of hc are seen to be quite strict. The z dependence represents a valuable check on possible spatial dependency.

A further restriction has been set by Colella et al (1975) who have devised and performed an experiment to test the influence of quantum effects and gravity as they involve h. They employed a Michelson interferometer, in which beams of neutrons interacted quantum mechanically. The measurements involved $g_E m_n/h$, where g_E is the local acceleration due to the Earth's gravitational field and m_n is the neutron mass. The results showed that h is independent of the gravitational field for weak fields, and verified the Principle of Equivalence in the quantum limit.

The electronic charge is similar to h, in that it has been suggested that it is cosmologically variable, but has had strict limits set on any such tendency. Gamow (1967a) was deterred by the increasingly stringent limits being set on a variation of G of the form $G \propto t^{-1}$, and advanced the alternative hypothesis that G is constant while $e^2 \propto t^{+1}$. This would cause the luminosity of the Sun to vary as $L_\odot \propto t^{-3}$, instead of the more drastic $L_\odot \propto t^{-7}$ resulting from the hypothesis $G \propto t^{-1}$. It would also lead to a redshift of uniform size for the lines of the spectra

118

of distant astronomical objects. It would further mean that isotopes which are unstable now would have been stable in the distant past. Immediately following the proposal of Gamow, three determinations of the time dependence of e were made (Dyson 1967, Peres 1967, Bahcall and Schmidt 1967), all of which came to the conclusion that it could not vary significantly, the best lower limit being

$$\frac{1}{e^2}\frac{d(e^2)}{dt} \lesssim 10^{-13} \text{ yr}^{-1}. \tag{4.44}$$

Broulik and Trefil (1971) have employed equation (4.44) in combination with a study of isotopes to derive a limit on the time variation of g_s, the strong coupling constant, which can be expressed as

$$\frac{1}{g_s^2}\frac{d(g_s^2)}{dt} \lesssim 2 \times 10^{-11} \text{ yr}^{-1}. \tag{4.45}$$

A similar limit, but more accurate by a factor of ten, has been obtained by Shlyakhter (1976), assuming that $dg_s/dt \simeq$ constant. His method, which utilises the resonance of the heavy-nucleus and slow-neutron system, also gives analogous limits on possible time variations of α and the weak coupling constant. These tend to show that neither parameter is varying on a 10^{10} yr time scale.

The theory of the strong coupling constant enables limits on the variability of other parameters to be obtained from limits on g_s. Incidental to the derivation of equation (4.45), Broulik and Trefil (1971) were thus able to set a close limit on the time variation of the fine-structure constant, α, from data on the isotope $^{244}Pu_{94}$. This limit complements the astrophysical one of equation (4.43) above, and is

$$\frac{1}{\alpha}\frac{d\alpha}{dt} \lesssim 2 \cdot 3 \times 10^{-11} \text{ yr}^{-1}. \tag{4.46}$$

Similarly, limits on the variation of g_s can be used to infer comparable limits on e, since if one is nearly constant then so must be the other. (A detailed account of nuclear theory, as it is affected by time variations of the coupling parameters, with refinements of the cosmological arguments for variability of

Dyson and Teller, has been given by Davies (1972).) It has been estimated by Baum and Florentin-Nielsen that $\dot{e}$ (and $\dot{m}_e$, $\dot{c}$, $\dot{h}$) must be essentially zero with regard to a possible variation on a time scale of the order of the age of the Universe. This they calculate on the basis of a limit $\Delta(hc)/(hc) \lesssim 10^{-12}$ yr^{-1} arrived at by comparing photons from near and remote astronomical sources (Baum and Florentin-Nielsen 1976, see also Solheim *et al* 1976). If QSOs are taken as being at cosmological distances, the limit on hc is of the order $\Delta(hc)/(hc) \lesssim 2 \times 10^{-14}$ yr^{-1}. These limits strongly suggest that the noted parameters, including g_s, are not noticeably time-dependent as judged at the present epoch. Dyson (1971) has pointed out that in any case, g_s and other parameters that control nucleosynthesis cannot have been much different from their present values in the early Universe, since otherwise the majority of the matter in the cosmos would not be observed at the present epoch to be unprocessed hydrogen. (The effect of the scalar–tensor theory on the helium abundance has been discussed by Dicke 1968.) The foregoing data suggest quite forcibly that G, via its probable connection with α, may indeed be a true constant, so far as can be judged from astrophysical evidence.

A related dimensionless number involving α, the masses of the electron (m_e) and proton (m_p) and the nuclear g-factor for the proton (g_p) can also be studied by astrophysical means and gives the limit

$$\frac{d}{dt}\left[\ln\left(\frac{\alpha^2 g_p m_e}{m_p}\right)\right] \lesssim 2 \times 10^{-14} \text{ yr}^{-1}. \tag{4.47}$$

This limit was obtained from observations of a QSO with coincident optical and radio redshifts (Roberts *et al* 1976, Wolfe *et al* 1976), the redshift of $z = 0.52385$ providing a good limit on the noted parameter.

With regard to the coupling constant for β decay, g_β, if G varies according to $G \propto t^{-1}$, then the weak coupling constant $g_w(\propto g_\beta^2)$ is expected to change at a rate proportional to t^{-n} (where $0.25 < n < 0.5$) and there should be a discrepancy between α and β decay ages. From meteorites, Dicke (1959a) found that there does not appear to be any evidence ruling out such a variation. If G were higher in the past, the radiation rate ($\propto G^7$ or G^8) and temperature of the Sun would have been higher, and so asteroids and meteorites in space would have been hotter. Potassium–argon ages can be employed on this basis to set a limit of $|\dot{G}/G| \lesssim 10^{-10}$ yr^{-1}. (Other data, of course, give closer limits, as discussed in Chapter 5.) Thus there is no evidence against a small variation of the coupling constants for the two weak interactions, β decay and gravity. This is in accordance with the viewpoint of the Brans–Dicke theory (Dicke 1962b, p36), that Mach's Principle can be properly incorporated into physics if the dimensionless numbers involving G are allowed to vary. On the other hand, there is no evidence *in favour* of the variability of these two coupling constants, while there is strong evidence *against* any variability of the strong and electromagnetic coupling constants and Planck's constant.

4.10. Constraints and Conspiracy

The dimensional numbers such as h, which are discussed in §4.9, have values depending on the system of units used to measure them. Changing yardsticks for metre rules is not a fundamental operation that needs to be incorporated into physical theory, and some workers have expressed serious doubts concerning the meaning of variability in parameters having nonzero dimensionality. It would seem, nevertheless, that Cavendish's experiment, carried out with an apparatus in good working order, could imaginably give values of G that might be found to vary secularly with time from $G \simeq 6 \times 10^{-8}$ cm^3 g^{-1} s^{-2} in such a way as to indicate some underlying non-trivial process. The Eddington numbers (§4.2), on the other hand, are dimensionless, and so the hypotheses (§§4.3–4.7) suggested as explanations of them can incorporate certain types of variability without the danger of epistemological paradox.

It is pertinent at this point to summarise results of previous sections on the constraints and expected variation rates of the parameters involved (G, h, c, α, g_s and e). One can anticipate a limit for $\dot{G}/G$ from the next chapter, and include limits on passive and inertial mass equivalence from the preceding one. The Eötvös experiment is upheld by numerous data, with a direct determination of $|\dot{G}/G| \lesssim 4 \times 10^{-10}$ yr^{-1}, and an indirect determination of $|\dot{G}/G| \lesssim 2 \times 10^{-11}$ yr^{-1} (see §5.8).

Also, $m_P(e^-) \simeq m_I(e^-)$ can be determined to one part in 10^4, and $m_P(e^-) \simeq m_P(e^+)$ to one part in 100. If $|\dot\alpha/\alpha| < 3 \times 10^{-13}$ yr^{-1}, then $|\dot G/G| \lesssim 2 \times 10^{-11}$ yr^{-1}; while $\Delta m_I/m_I < 10^{-20}$ for inertial anisotropy. Further, $\mathrm{d}[\ln (hc)]/\mathrm{d}z \lesssim 3 \times 10^{-4}$; $\mathrm{d}[\ln (hc)]/\mathrm{d}t \lesssim 10^{-10}$ yr^{-1}; $\mathrm{d}[\ln (\alpha^2)]/\mathrm{d}t \lesssim 10^{-11}$ yr^{-1}; $e^{-2} \, \mathrm{d}(e^2)/\mathrm{d}t \lesssim 10^{-13}$ yr^{-1}; $g_s^{-2} \, \mathrm{d}(g_s^2)/\mathrm{d}t \lesssim 2 \times 10^{-11} - 2 \times 10^{-12}$ yr^{-1}; $\alpha^{-1} \, \mathrm{d}\alpha/\mathrm{d}t \lesssim 2 \cdot 3 \times 10^{-11}$ yr^{-1}. Most of these represent direct-measurement limits, but other methods which depend on reasonable assumptions (Dyson 1967, 1972a, b, Davies 1972, Shlyakhter 1976) give $\alpha^{-1} \, \mathrm{d}\alpha/\mathrm{d}t \lesssim 10^{-17}$ yr^{-1}, $g_s^{-1} \, \mathrm{d}g_s/\mathrm{d}t \lesssim 5 \times 10^{-19}$, $g_w^{-1} \, \mathrm{d}g_w/\mathrm{d}t \lesssim 2 \times 10^{-12}$ yr^{-1}.

There are also limits on the ratio of active to passive gravitational mass in fluorine and bromine of 5×10^{-5} (Will 1976; §3.6); and of the equality of m_I and m_P, as they might be affected by the energy of weak interactions, of at least one part in 100 in comparison with the rest mass energies as used in the Eötvös experiment (Haugan and Will 1976). The combination of numbers given in equation (4.47) is probably the most accurate available limit on the time variability of the parameters it involves. There is, as yet, no evidence of variability of the weak coupling constant, g_w, with time, but there is extant one determination (Chapter 5) of a finite rate for $\dot G/G$ that might imply a time variability for g_w and, perhaps, for some of the other parameters mentioned above.

It is conceivable that the dimensionless numbers, such as the Eddington number $Gm_p^2/\hbar c$, could be absolute constants, while the component parameters comprising them are variable. The possibility of such a conspiracy between the dimensional numbers was recognised by Dirac (see Harwit 1971). The admissibility of this conspiracy hypothesis—as it has been called—can be judged only when a large amount of data is available on the variability of the familiar constants of physics.

5.   *The Rotation of the Earth*

5.1. Introduction

The spin of the Earth on its axis is not the most accurate of time-keeping devices, since atomic clocks can monitor changes in the rate of rotation of the planet over a spectrum of time scales. It does, however, provide the system with the largest size and the longest run of recording time from which reasonable accuracy can be expected. The Moon raises tides in, and on, the Earth which, due to friction, lag behind the Moon as it moves in its orbit. The energy lost in the tides appears as a secular acceleration of the Moon's position, and as the Moon retreats from the Earth, the Earth itself spins down by an even greater amount. The combined system is large enough to manifest the possible presence of those cosmological processes that have effects of the same order of magnitude as the geophysical ones, and so are amenable to observation.

5.2. Rotational Perturbations

The theory of the rotation of the Earth is a complicated subject and has ramifications in many attendant fields of study. Cosmological effects on the angular velocity of rotation of the Earth (ω) and its rate of change ($\dot{\omega}$) are closely connected with changes in the angular velocity of the Moon in its orbit (n) and its angular acceleration ($\dot{n}$). To appreciate the relevance of cosmological processes, it is necessary to have some understanding of the competing geophysical effects, both as to size and origin.

The spin axis of the Earth has an inertial orientation that changes with a period of 26 000 yr (the steady precession, amplitude $\simeq 23°$). General relativistic effects contribute $\simeq 2''$ century^{-1} to this precession. Other periods are 18·6 yr (the principal nutation, amplitude $\simeq 9''$), 9·3 yr, 1 yr, 6 months, 2 weeks (parts of the nutation in obliquity and longitude, amplitude $\lesssim 1''$), and possible other changes due to ill-understood, perhaps general relativistic effects (secular decrease in the obliquity with a discrepant rate $\approx 0''.1$ century^{-1}). The spin axis also moves with respect to the surface of the Earth, and has components with periods of 20–40 yr (a perturbation of amplitude $\approx 0''.01$, sometimes referred to as the Markowitz wobble); 440 days (the Chandler wobble, amplitude $\simeq 0''.15$ but variable, damping period $\simeq 30$ yr); 1 yr and 6 months (seasonal perturbations of amplitudes of $0''.10$ and $0''.01$, respectively); 1 month and 2 weeks (amplitude expected to be $\approx 0''.001$); 1 sidereal day (diurnal, period $\lesssim 0''.01$); other changes with the nutational periods (Oppolzer terms, amplitude $\approx 0''.01$); longer-term, secular motions are irregular (rate $\approx 0''.1$ century^{-1}).

The instantaneous axis of spin defines the angular velocity (ω) of the planet, and this changes over periods of 10^{10} yr (secular acceleration, $\dot{\omega}/\omega \simeq -5 \times 10^{-10}$ yr^{-1}); 100 yr ($\dot{\omega}/\omega \lesssim \pm 5 \times 10^{-10}$ yr^{-1}); 1–10 yr ($\dot{\omega}/\omega \lesssim \pm 100 \times 10^{-10}$ yr^{-1}); months ($\dot{\omega}/\omega \lesssim \pm 500 \times 10^{-10}$ yr^{-1}), and other short-period changes, some of which are doubtful (2 yr, $\Delta\omega \simeq 10$ ms; 1 yr, $\Delta\omega \simeq 20$ ms; 6 months, $\Delta\omega \simeq 10$ ms; 1 month and 2 weeks, $\Delta\omega \simeq 1$ ms). These movements and changes relating to spin can be measured at present to an accuracy of $\pm 0''.01$ over periods $\simeq 6$ months, but a potential accuracy of order $0''.001$ is offered by the technique of very long baseline interferometry using radio sources. As noted by Rochester (1973), any divergences that turn up in the future between Ephemeris Time and the Atomic Time service will indicate inadequacies in the theory of the Moon or in Newtonian celestial mechanics.

Since the data became accurate enough to detect possible cosmological effects, several results concerning the movement of the axis and changes in ω have been obtained (Rochester 1973, pp774–7). It has been calculated that anisotropies in G predicted by preferred-frame theories of gravity (§3.7) could have noticeable effects on the rotation. The effect of winds in producing abrupt changes in the length of the day with periods $\simeq 1$ week have, however, been attributed more importance. There has also been partial confirmation by geochronology of the older derived values for $\dot{n}$, obtained by Spencer Jones and others (see below). These subjects are discussed by Rochester (1973), but the problems are far from settled. The study of the Earth's rotation is undergoing scrutiny, largely because it is now realised that more processes are of possible importance with regard to changes in ω and $\dot{\omega}$. These include effects of the solid inner core of the Earth; the inertial, viscous, turbulent, electromagnetic and topographic coupling across the core–mantle boundary (CMB); the elasticity, dissipation, viscosity and seismic state of the mantle; the dissipation, loading and inertia associated with the oceans; the loading and inertia of groundwater; the winds, tides and inertia of the atmosphere; the lunar gravitational couple; the solar gravitational couple, and the solar wind or interplanetary torque. Cosmological effects on ω and n act against a background comprising these and related geophysical effects. The main predicted process is the secular spin-down of the planet (due to a decreasing G, or global expansion) and its consequences with regard to changes in the constitution and shape of the Earth.

5.3. Angular Momentum in the Earth–Moon System

If the influence of the Sun is neglected, the Earth–Moon system can be considered as a classical two-body problem in which angular momentum is exchanged between the rotational and orbital motions. There has long been a controversy about the history of the Moon's orbit and the interaction of the Moon with the Earth in earlier geological epochs. This controversy centres around discrepancies between different methods of determining the secular accelerations involved (Jeffreys 1970). The palaeontological method (see §5.4) gives the longest run of data, while more recent information can be obtained from ancient eclipse observations, telescope observations and occultation data based on the Atomic Time service. The latter are treated in §5.8.

The history of angular momentum transfer within the Earth–Moon system must be closely connected with the results of non-Newtonian gravitational theory, if such theories are admissible. For the case of a G dependence like that of the cosmologies considered previously, Vinti (1974) has examined the evolution of a two-body system in which $G \propto (\text{constant} + t)^{-1}$, the constant being introduced to avoid embarrassment near $t = 0$. If the semi-major axis of the orbit is l, n_m is the mean motion of the secondary about the primary, and P is the period, then Vinti shows that

$$\frac{\dot{G}}{G} = -\frac{\dot{l}}{l} = \frac{\dot{n}_m}{2n_m} = -\frac{\dot{P}}{2P}. \qquad (5.1)$$

The object of much work in geophysics is to determine $\dot{n}_m/n_m$ as it applies to the Moon. Unfortunately, $\dot{n}_m/n_m$ so found is not directly usable in equation (5.1) because the Earth–Moon system is non-conservative and dissipates energy by tides, etc. Most of the observed $\dot{n}_m/n_m$ is due to dissipative processes, and the contributions to $\dot{n}_m/n_m$ from geophysical effects must be removed before $\dot{G}/G$ can be calculated from equation (5.1). Unless otherwise specified, $\dot{n}_m/n_m$ will be taken to mean the total deceleration rate. In practice, observations to determine $\dot{n}_m/n_m$ for the Moon are made from the Earth, which itself is being de-spun by friction and would also be affected by a decreasing value of G. Cosmological theories in which G decreases with time lead to the inevitable result that the Earth should be expanding. There may be independent evidence in favour of expansion, unrelated to the variable-G theories (Wesson 1973). As the Earth expands, its moment of inertia increases and its spin decreases, if no other forces are acting. In general, the actual situation is complicated because neither $I\omega^2$ nor $I\omega$ (Runcorn 1968a, b) is constant for the Earth (I is the moment of inertia). Any expansion effect is likely, in any case, to be masked by the slowing due to tidal friction (Jeffreys 1970, pp309–13), the total increase in the length of the day (LOD) being about 2 s in 10^5 yr. This rate has probably not been constant, since the palaeontological data of Pannella et al (1968) show a notable cessation of the spin-down in the Palaeozoic. The same data have been used to investigate the history of the Moon's orbit.

At present, the Moon is believed to be receding from the Earth at $\simeq 3$ cm yr^{-1}, and its angular velocity is consequently decreasing (Scrutton and Hipkin 1973). This is inferred from measurements of the angular acceleration of the Moon's position in orbit around the Earth from three different sources: (i) telescope observations give an angular acceleration of

126

−22″.4 (±1·0) century^{-2} (Spencer Jones); (ii) eclipse observations give −37″.7 (±4·4) century^{-2} (de Sitter), −41″.6 (±4·3) century^{-2} (Newton) and −34″.2 (±1·9) century^{-2} (Stephenson); (iii) star occultations give −52″ (±16) century^{-2} (Van Flandern), −42″ (±6) century^{-2} (Morrison) and −38″ (±8) century^{-2} (Oesterwinter and Cohen). If the angular momentum of the Earth–Moon system is constant, these data imply that the LOD is increasing at about 2·4 ms century^{-1} (using de Sitter's result as an illustration). This is comparable with the 2 ms century^{-1} increase in the LOD for secular time scales inferred from palaeontology. Even though the angular momentum of the Earth–Moon system is not really constant, the discrepancies in the above data are notable, and lead one to suspect that some process is at present trying to accelerate the Earth. This could be the accretion of material by the core, a tectonic effect, or (most probably) coupling across the core–mantle boundary. The size of the accelerating term is large, because since the year −762 the change $\dot{\omega}/\omega$ in the Earth's spin has averaged about −25 parts in 10^9 century^{-1}, while the part not attributable to tides has averaged about +23 parts in 10^9 century^{-1} (Newton 1972b). This does not mean, of course, that the accelerating term has always been this large, since it could easily be a temporary phenomenon. One explanation has been given by Yukutake (1972), who has shown that an oscillation of the magnetic dipole moment of the Earth can change the angular velocity of the mantle by considerable amounts, the predicted sense being in agreement with a spin-up term at the present stage of the cycle.

A further discrepancy in the rotational theory, that of the excess secular change in the obliquity of the ecliptic noticed by Aoki and discussed in Wesson (1973), was formerly thought to imply a large secular deceleration of the Earth's spin. Aoki (1967) proposed a connection between the dynamics of the Earth's core and the Galaxy to account for an unexplained secular change in the obliquity of the ecliptic that may have been due to some cause involving the Earth's equatorial motion. If the Earth were perfectly rigid there would be no hope of such an explanation, but Aoki suggested that the core and mantle may have differently directed angular velocity vectors, the core being considered fluid. The secular change was used by Aoki to infer a revised value for Oort's constant (B) which is determined by observing systematic trends in the proper motions of certain stars. This revised value implied that the mass of the Galaxy needed to be increased by a factor of 2·8 over the generally accepted value. Sekiguchi (1967) further investigated the proposal of Aoki concerning the core and mantle, using the assumption that the rotation vectors of core and mantle are equal in modulus. His analysis indicated that large torques might be operating in the core. What the effect of such couples would be on the generation of magnetic fields in the core is difficult to say, but a maximum entropy power spectrum analysis of long-period geomagnetic reversals by Ulrych (1972) showed that periodicities of 700 ($\pm$ 100) $\times 10^6$ yr and 250 ($\pm$ 30) $\times 10^6$ yr exist in the reversal pattern, these presumably having some relation to large-scale phenomena affecting the core.

Some doubt was always present about Aoki's original hypothesis, and the very large secular change in the Earth's spin which, as an explanation of the ecliptic anomaly, it entailed. Aoki's interpretation depended somewhat on observations of stars and the structure of the Galaxy (Fricke 1972). The anomaly has been, in part, resolved as largely non-physical following re-analysis using a better model of the Earth by Kakuta and Aoki (1972). There are numerous other puzzling data in need of explanation (Newton 1968a, b, 1969, Jeffreys 1970, pp304–16, Dicke 1969, Challinor 1971, Maran and Ögelman 1969, Pfleiderer 1970, Hipkin 1970a), none of which is in definite contradiction to $\dot{G}$ theories, and some of which might be resolved if G were time variable.

Claims that an atmospheric tide, of the sort first noted by Kelvin, could be keeping the length of the day constant through long periods of time, are almost certainly wrong. (Such claims have been made by Holmes (1965, p972) and Öpik (1970) and these have been criticised by Jeffreys (1970, p307) and Asnani (1969).) It is obvious from the plot of angular momentum density (in rad cm^2 s^{-1}) against the mass of the planets forming the Solar System (MacDonald 1966) that the Earth, Venus and Mercury are all deficient in angular momentum and lie nearest the Sun (which induces tides in these bodies), while Uranus, Neptune, Saturn and Jupiter form a roughly linear series in this plot with higher angular momentum density. Further-more, the solar thermal couple on the Earth's atmosphere is measured to be only a quarter of the solar couple acting on the world's seas (Dicke 1966), which is itself small. These two results are both in agreement with the observation by MacDonald that the angular momentum density, A_{pl}, is correlated with the mass, M_{pl}, of a planet by $A_{pl} \propto M_{pl}^{0.8}$. ($A_{pl} \propto R_{pl}^2 \omega_{pl}$, where R_{pl} and ω_{pl} are the radius and angular velocity of the planet, respectively.) This reflects the fact that bodies (including asteroids) in the Solar System rotate with periods all within a factor of three of 10 h, although the mass range is a factor of 10^{12} (see Hartman and Larson 1967). Such a relationship must reflect the mode of origin of the planets (Fish 1967). Thus they may have formed by a mechanism similar to that proposed by Alfvén, in which a plasma rotating about the proto-Sun fed the accretion of bodies that formed with rotational periods independent of their sizes (Alfvén 1964b). It is also consistent with a similar accretion process for the asteroids (Alfvén 1964a). The observed independence of the periods and radii might also be the result of some process in which bodies were rotating at their stability limits. The stability limit hypothesis, if true, would mean (Fish 1967) that all bodies in the Solar System have had their rotational velocities decreased by about one-third over geological time, independent of their distance from the Sun. Cosmological theories would be possible candidates as possible explanations of such a secular loss of angular momentum in the planets.

Weinstein and Keeney (1973a) have re-examined the data of Pannella *et al* with a view to estimating the apparent loss of angular momentum in the Earth–Moon system. They find that there has been an overall 5% loss, with the Moon having lost angular momentum over the period Pre-Cambrian to Silurian, having been constant in the interval Silurian to Pennsylvanian, and having gained angular momentum from Pennsylvanian to Recent. The first two phases are somewhat anomalous, since tidal friction can only increase the lunar momentum. The length of the sidereal year has been approximately constant over the last 10^9 yr at least (Hipkin 1970b), so the implication is that either G is changing, or I is changing. If $\dot{G} \neq 0$, the data require that G has *increased* secularly: this harks back to Milne, but is hardly satisfactory. The alternative is that I has increased by 25–30% over the last $1 \cdot 7 \times 10^9$ yr, and that a large part of the increase occurred before the Devonian. This may support an expanding Earth model, but, as noted by Weinstein and Keeney, the anomalous behaviour of the lunar momentum over the period Pre-Cambrian to Pennsylvanian cannot be entirely due to changes in I. A fairly good case could be made for saying that part of the effects found are due to changes in the orbital eccentricity of the Moon. If the shape, or orientation, of the lunar orbit changes, or if the tilt of the Earth's axis in space alters, angular momentum is exchanged in such a way as to give an apparent degree of non-conservation (Hipkin 1975). If only data on the length of the day and month are considered, the angular momentum of the observed lunar acceleration is greater than that apparently lost due to the observed slowing of the Earth's rotation. Hipkin (1975), besides drawing attention to this point, also gives the basic equations from which the effects on the rotation and local acceleration of gravity, due to changes in the moment of inertia and the Earth's radius and ellipticity, can be calculated. He shows that changes in the lunar eccentricity, lunar inclination and the obliquity of the ecliptic can all cause relative angular momenta changes of varying amounts, as well as the Earth's rotational perturbations. (Newton (1975) has shown that the obliquity of the ecliptic 2000 years ago is in agreement with the value predicted from Newcomb's theory.) Hipkin, while also drawing attention to the possibility that the solar atmospheric tidal torque, although small over historical times, may nevertheless have been an important source of spin-up in the Earth long ago, has reached two conclusions of significance. (*a*) The data on deceleration do not allow one to detect empirically a net gain in the Moon's orbital angular momentum, or a net loss from the Earth's rotation, and (*b*) the data show no empirical support for the widespread belief that the size of the lunar orbit has not varied by an amount greater than that predicted by conservative gravitational theory.

Point (*b*) means that the lunar orbit could be expanding because of cosmological effects (Chapters 6 and 7). Point (*a*) is closely connected with the work of Weinstein and Keeney (1975), who have plotted the palaeontological data of Berry

130

and Barker, Pannella *et al*, Mazullo and Pannella, as discussed in §5.4. A plot of the number of days in a year against the number of days in a synodic month follows a line that corresponds, somewhat surprisingly, to a relation in which this ratio is a constant, and in which there is no tidal interaction and a constant lunar eccentricity. It could well be, however, that this relation, which is based on a small range of data, is also connected with other interpretations in which there is secular expansion and a variable lunar eccentricity. While the data used in this work are mostly younger than Palaeozoic, one result does emerge unambiguously from the palaeontological clock evidence. The Moon has been in orbit near the Earth for at least 2×10^9 yr (Pannella 1972), and the Earth has been decelerating for at least this long. This deduction is not easy to fit into the preceding discussion, but it agrees with the classical picture.

Weinstein and Keeney (1973b, 1974) have attempted to provide a cosmological explanation for at least part of the angular momentum loss by generalising Hubble's law. They note that the rest mass of the photon is usually taken to be zero, but that experimental limits and a consideration of Proca's electrodynamic equation admit a finite value in the range 10^{-43}–10^{-56} g. Weinstein and Keeney therefore consider a tired-light cosmology, in which the rest mass of the photon turns out to be less than 10^{-63} g, and in which Hubble's law is written as

$$\frac{d\nu_p}{dl_p} = -\frac{H\nu_p}{c},$$

where H is Hubble's parameter, c is the velocity of light, ν_p is the frequency of a photon and l_p is the path length travelled by the photon. In so far as tired-light cosmology is of doubtful, but not completely disprovable, validity (Peebles 1971), this conjecture by Weinstein and Keeney is admissible. The hypothesis is open to test since it predicts a general angular momentum loss, as can be seen by multiplying the last equation by Planck's constant, h, when the energy equation

$$\frac{d(h\nu_p)}{dl_p} = \frac{d(h\nu_p/c)}{d(l_p/c)} = -H\left(\frac{h\nu_p}{c}\right)$$

is seen to be equivalent to a change in the momentum of a

propagating photon at a rate $dp_p/dt = -Hp_p$. The drag force $(-Hp_p)$ is, in fact, a general phenomenon that would act on all material systems of astrophysical and geophysical type. The observed 5 % loss in the angular momentum of the Earth–Moon system over the last $1\cdot7 \times 10^9$ yr can be used to calculate Hubble's parameter, from the hypothesis, as $H \simeq 4 \times 10^{-18}$ s^{-1}, in approximate agreement with astronomical observation. The outlined hypothesis is one possible way to account for angular momentum loss, but it is difficult to isolate any such one effect from the many contributing influences of changes in the Earth's shape and surface topography. It is particularly difficult to estimate the effects of tides on the slowing of the Earth over long periods of time, and this uncertainty leaves open the question of there being a notable contribution to the spin-down from G variability, global expansion, or other cosmological effects.

Jeffreys has re-analysed the tidal friction data, but assumes that the angular momentum of the Earth–Moon system and the moment of inertia of the Earth have remained constant (Jeffreys 1973). These are slightly erroneous conditions, considered in the light of the work of Weinstein and Keeney, but Jeffreys' conclusions (that the secular acceleration of the Moon's mean motion has probably not altered in the last 2000 years, but that the secular acceleration of the Earth's rotation probably has changed) are still likely to be correct. Tidal friction theory is only accurate to a factor of two, so the spin-down could be due entirely to tidal friction. The variabilities that are encountered in eclipse observations, telescope observations and palaeontological data might then be results of changes in the configuration of shallow seas.

One thus has to conclude that the secular slowing of the Earth's rotation and the relaxation of the lunar orbit do not at a first examination provide any firm data for, or against, the operation of cosmological effects in the Earth–Moon system. The uncertainties of tidal friction analysis may mask the operation of cosmological processes leading directly to a loss of angular momentum, and may also mask the secondary processes (such as expansion) that can be predicted from certain cosmologies. It has been estimated by Birch (1967) that a change from $2G$ to G alone would cause an expansion of

370 km or so in the Earth. This estimate approximately agrees with those made by proponents of the Brans–Dicke and Hoyle–Narlikar cosmologies (Chapter 2). A further discussion of the expanding Earth hypothesis (Dooley 1973) is deferred until Chapter 6. Attention will be focused meanwhile on a closer analysis of possible cosmological effects on the rotation, with a view to predicting the sizes of such effects and delineating limits on them.

5.4. Palaeontological Data

The rate of slowing of the Earth's rotation of about 2 s in 10^5 yr, disclosed by historical and modern observations, is a process that has been affecting the Earth throughout geological time (as shown by geochronological data supplied by palaeontology). Corals especially, since they reflect fairly accurately the alternation of day and night, have been extensively employed to derive information on the Earth's rotation in past geological epochs.

In the following, an account is given of six groups of palaeontological data and their consistency with each other, and then the relationship of these results to connected work is examined. The six groups of data are as follows.

(*a*) Wells (1963) found that there were 400 days in the Middle Devonian year, while Newton (1969), on Wells' data, found 410. The figure of 400 seems to be indicated by the specimens employed. This is a difference of 35 from the present 365 days in the year, although it must be remembered that it is the length of the day which has changed, so a direct comparison of this kind is only an average figure.

(*b*) Scrutton (1965) gives 30·59 days in a synodic month for the Middle Devonian.

(*c*) Pannella *et al* (1968) have provided numerous data from which, for the purposes of comparison, one can take 29·92 days in the Upper Cretaceous synodic month (the age of this datum is given as 72×10^6 yr, the quoted standard deviation as $\pm 1·00$, and the standard error as $\pm 0·10$). Another figure is 30·53 days in the Middle Devonian synodic month (datum age $= 380 \times 10^6$ yr, standard deviation $= \pm 1·25$, standard error $= \pm 0·23$).

(d) Berry and Barker (1968) find 29·63 days in the Upper Cretaceous synodic month.

(e) Mazzullo (1971) has given results that agree approximately with those of group (c) above.

(f) Pannella (1972) has also given results agreeing with those in group (c) that extend the range of strata examined.

One notices that of the results (b), (c) and (d), the Cretaceous datum of (d) agrees with the Cretaceous datum of (c) within the error of the latter. Also, the Devonian datum of (b) agrees with the Devonian datum of (c) within the error of the latter. Since (d) and (c) are compatible (Cretaceous) and (b) and (c) are compatible (Devonian), I take it that (b) and (d) are also compatible (that is, the consistency of the methods used means that data of all ages in (b), (c) and (d) are compatible).

Are (b), (c) and (d) compatible with (a)? Berry and Barker claimed that their result gives 370 days in the Upper Cretaceous year (epoch $\simeq 70 \times 10^6$ yr), while the Wells (1963) figure is 400 days for the Devonian year (epoch $\simeq 400 \times 10^6$ yr). A rough calculation suggests that the $(370–365) = 5$ days difference of Berry and Barker would be compatible with the $(400–365) = 35$ days of Wells, since the ages for the two data are in the ratio $400/70 \simeq 6$. The agreement is not as good if the correction suggested by Runcorn (1968b, 1969b) is included, and the cessation of slowing in the Cretaceous really needs closer examination. The quantitative shape of the curve of Pannella et al (1968) is of importance because the data of Berry and Barker sampled an epoch (Upper Cretaceous) that came at the end of a long span of time (Upper Carboniferous→Upper Cretaceous) when the number of days in the synodic month was approximately constant. This may explain why the figure found by Berry and Barker is not in exact agreement with that expected from Wells' result. The cause of the cessation of the increase in the number of days in the synodic month could be found in a resonance effect (see below), or in a temporary change of the tidal friction regime, or in some other phenomenon. Tidal processes, which affect the length of the month, may have temporarily ceased to be effective due to some geological event, perhaps connected with the onset of continental drift. The length of the day, meanwhile, could still have been increasing due to G variability or expansion of the globe, but would have had no measurable result on the palaeontological data now available for examination.

While the above results seem to indicate a rough consensus that the Earth has been decelerating since the Palaeozoic at about 2 ms century^{-1}, as judged from palaeontological studies, some level of caution is advisable in interpreting these data. Firstly, dating of the samples is sometimes controversial (for example, Hazel and Waller (1969) criticised the dating procedures of Pannella *et al*, although the criticism has been effectively replied to), and other aspects of the palaeontological method have been given a critical appraisal by Scrutton and Hipkin (1973). Secondly, the correction factor of Runcorn (1968b, 1969b) mentioned above is sometimes overlooked. The latter is simply a result of Kepler's laws, and says that the fractional change in the length of the day is 0·40 or 0·43 (depending on the law of tidal friction) of the fractional change in the number of days in a synodic month. Thirdly, there are two other direct ways of estimating the length of the day in the past that do not always agree with the results of palaeontological studies.

These two alternative ways are those involving telescopic observations since the seventeenth century and ancient eclipse data. The eclipse data, on some interpretations, give a rate of spin-down of the Earth that is twice as great as the rate indicated by the telescopic or coral data (Runcorn 1964). Eclipse records are well known for the fact that observations from ancient times often appear to be unreliable. This suggests that the palaeontological and telescope data should be given greater credence than the eclipse data. Runcorn (1964, 1968a) has calculated on this basis that the spin-down results do not favour expansion of the Earth at rates of about 1 mm yr^{-1}, as suggested by Egyed and Creer, unless G is also changing (since the lunar orbit is then modified as well as the Earth's spin rate). This depends to a certain extent on the assumptions made as to whether iron is still draining into the Earth's core from the mantle, but it would appear to be true from coral data that the torque responsible for the slowing of the Earth's rotation is approximately the same now as it was in the Devonian. Most of the loss of angular momentum represented by this torque is connected with tidal friction.

While the average acceleration of the Earth over the last 2000 years has been negative overall, a careful re-analysis of the eclipse observations indicates that there is some non-frictional torque operating that is trying to accelerate the Earth's spin (Newton 1969). This could be a geophysical process internal to the planet, and will be mentioned again below. Newton (1968a, 1969) finds that the overall deceleration *could* contain a part due to a varying G of the size advocated by Dicke (1969), but such a contribution is not necessary. A term due to G might be detected in principle by using observations of the period of precession of the nodes of the lunar orbit (Runcorn 1969a). It is difficult to rule out $\dot{G}$ effects, because ambiguities exist in the methods used to reduce the data. The latter concern the precise type of tidal friction operating (Newton 1968a); the contending influences of gravity-induced and solar tides (the two together provide about 17–21 parts in 10^{11} yr^{-1} in deceleration, in rough agreement with a careful analysis of eclipse observations); short-term changes in the secular accelerations of the Sun, Moon and planets due to processes interior to the Earth (such as electromagnetic coupling changes across the CMB; Jeffreys 1970, p305) or upper atmosphere super-rotation effects (King-Hele 1970); and other short-term fluctuations in the spin rate and problems with the way in which these are analysed (Maran and Ögelman 1969, Pfleiderer 1970, Challinor 1971). These uncertainties in effects which are involved implicitly in spin-down data are further compounded by the possibility of large-scale geophysical movements which seem to be indicated by certain palaeontological data (see below). It should nevertheless be emphasised that, while discrepancies may exist between methods of determining the deceleration (or discrepancies within one group of data indicating times of anomalous spin-down), it is virtually beyond dispute that most of the spin-down is present and continues at a steady rate because of tidal friction (see, for example, Jeffreys 1970). Palaeontological data are in overall agreement with tidal friction estimates. Jeffreys (1975), in reviewing the tidal friction data, estimated that coral data give a change in ω since the Devonian (370×10^6 years ago) of 10%

and a change since the Cretaceous (60×10^6 years ago) of 1·5%. The theoretical values calculated from modern tidal data are about 10% and 2·5% so that a discrepancy may exist for more recent geological epochs, if not for the Palaeozoic.

The importance of geochronology involving synodic month intervals lies in the ability it has to separate the influence of tidal friction (which affects the length of the day and the lunar orbit) and changes in the moment of inertia of the Earth (which only affect the length of the day). Unfortunately, large changes in the Earth's rotation rate and in the Moon's orbital period result in only a small change in the synodic month, and matters are complicated further by an expected resonance involving the two effects, which might have occurred when the year contained 390 days and exactly 13 synodic months. This resonance may account for the interval (Pennsylvanian to Cretaceous) during which the results of Pannella *et al* indicate a pause in the lengthening of the day (Miller 1969), but such an interpretation is by no means certain.

The changes in rate of spin-down as deduced from the data of Pannella *et al* (1968) have been interpreted geophysically by Creer (1975), who has also inferred on rather weak grounds that discontinuities exist in other palaeontological data. He has attempted to correlate these changes with discontinuities in the integrated percentage of reversed magnetisation as compared with normal magnetisation in rock samples. Discontinuities in the latter occurred 425, 375, 220 and 55 million years ago. Discontinuities in the number of days in the month may have been present at past epochs of 420, 360, 215 and 50 million years, and in the number of days in the year at past epochs of 415, 360, 245 and 65 million years. Creer considers the following in trying to find an explanation for the claimed correlations.

(i) Thermal expansion of the mantle (unacceptable, because a temperature change of 2000 K would have been needed to produce the 0·5 ms century^{-1} change in the length of the day for the Mesozoic).

(ii) Temperature-induced phase changes in the mantle (reasonable, since effects of up to 2·7 ms century^{-1} over 100×10^6 yr with a radius decrease of 100 km could be produced if the upper 600 km of the mantle changed its rock structure to the lower mantle form with a 20% increase in density).

(iii) Changes in the moment of inertia, or ellipticity, of the Earth (also feasible, since changes in the moment of inertia are in about the same proportion as the differences in the hydrostatic and actual ellipticities, that is, 1/299·7 against 1/298·3 or a 0·5% difference, and the alteration of the difference by an amount of this order could instigate an effect of size 1 ms century^{-1} over a time scale of 100×10^6 yr). The last mechanism is connected with the argument of Goldreich and Toomre that a non-hydrostatic ellipticity of 0·5% is to be expected in a randomly evolving spheroid such as the Earth, and so should not necessarily be judged as a spin-down phenomenon.

Creer (1975) has also considered the displacement of the geomagnetic dipole from the centre of the Earth and extra-terrestrial mechanisms. The latter include a possible connection with the Fourier power spectrum for palaeomagnetic reversals as calculated by Crain et al (1969), which showed two major components with periods of 300 and 80 million years. These probably do not have a terrestrial source (Crain and Crain 1970), but may be linked with the Earth's position in the Galaxy and the vibration of the Solar System perpendicular to the Galactic plane, which have periods of 280×10^6 yr (orbital) and 84×10^6 yr.

While the correlation of the changes in spin-down rate, as indicated by Creer from palaeontological data, might be explicable as the effect of phase changes, ellipticity or inertia changes and possibly more far-reaching mechanisms, it might also be related to fluid dynamics inside the core. Jacobs and Aldridge (1975) have analysed the angular velocity of a fluid in a cavity whose walls are undergoing spin-down, and have found that a steady state lag might be set up in the motion of the mantle and the fluid in the Earth's core. This produces a relative drift of 0·5 mm s^{-1}, if the viscosity of the core material is $\nu_c \approx 10^{-2}$ cm^2 s^{-1} (10^{-6} m^2 s^{-1}) as indicated by geophysical considerations. This drift speed, caused by the material of the core lagging by viscous coupling behind the mantle spin rate, is approximately equal to that associated with the westward drift of the geomagnetic field. Changes in the configuration of the core, as advocated by Runcorn (1975) and others, could presumably provide a connection between palaeo-magnetism and rotation in the sense of Creer's correlations. Runcorn, besides drawing attention to the weak bases of Creer's palaeontological inferences, also disagrees with the neglect by Jacobs and Aldridge of electromagnetic coupling across the CMB. Runcorn is of the opinion that, while Rochester showed that torques across the CMB fail by a factor of five to account for the decade fluctuations in the length of the day as calculated from data on the geomagnetic field and its secular variation, alternative data still make this mechanism a feasible one. This subject, discussed elsewhere in this chapter, is a complex one, and raises the possibility of the shorter period changes in the spin rate and the Chandler wobble being due to turbulence in the core coupling to the mantle. A mechanism of this kind is likely since the Eckman number ($E \equiv \nu_c / \omega_c r_c^2$, where ν_c is the kinematic viscosity of the core, ω_c is the angular velocity and r_c is a typical length parallel to the axis of rotation), which is a measure of the ratio of viscous to Coriolis forces, is only of very small order ($E \ll 10^{-5}$) in the Earth's core (Jacobs and Aldridge 1975, p346). The core is therefore a place for the origin of several processes which, via the geomagnetic field and direct coupling in accordance with the conservation laws, could result in angular momentum transfer that would eventually alter the angular momentum balance of the Earth's core and mantle, and so of the Earth–Moon system.

A further indication that the angular momentum of the Earth and the Moon in the distant past may have been varying in a way not now seen is provided by studies of the lunar orbit (Jeffreys 1970). Palaeontological data on the length of the day, assuming that the length of the year has not altered, can be used to investigate the lunar history, and these suggest that the Moon must have been close to the Earth, at the Roche limit, about $2 \cdot 8 \times 10^9$ years ago (Turcotte et al 1974). Estimates of the epoch at which the Moon was near the Earth vary, but all raise the problem of where the Moon was before that epoch, and what laws of spin-down are necessary to allow a consistent lunar orbital history to be developed. If the Moon was at its Roche limit $2–3 \times 10^9$ years ago, this could be connected with

Pre-Cambrian volcanism and continental drift. When the length of the day was shorter, the annual exciting period of the Chandler polar motion must have passed through a resonance with the 14 month (428 day) wobble, provided that the length of the year did not also alter. This idea by Cannon (1974) and related views have been summarised by Hughes (1974). The resonance, if it occurred, would have caused the deposition of about 10^{26} erg yr^{-1} (10^{19} J yr^{-1}) in the Earth over a period of 5–25×10^6 yr about 200×10^6 years ago. The consequent rise in temperature of the asthenosphere by 1–10 K would have decreased its viscosity by 5–50% and this could have triggered continental drift. The dragging of lithospheric plates by tidal forces has been considered by Bostrom *et al* (1974) and Jordan (1974), and while it would seem that the couple is not strong enough to drive plate motions at the present time, it may have been in the past.

Stephenson (1972; see also §6.6) has re-examined all the data available, discussing the geophysical, astronomical and chronological inferences of the whole subject, and sums up the situation as follows. Telescope observations of the sort used by Spencer-Jones (see Munk and MacDonald 1960) give a rate of LOD change of 1·6 ms century^{-1}. This is not in great disagreement with the most careful recent analyses of eclipse observations, which give 2·4 ms century^{-1} (Newton 1969, 1972a) or 2·2 ms century^{-1} (Stephenson 1972). These in turn are in approximate agreement with palaeontological data such as those of Wells (1963), which all give about 2 ms century^{-1}. These are all net decelerations, and there is evidence that, at least for relatively recent times, the average rate of change of LOD of -2 ms century^{-1} has been composed, loosely speaking, of two parts. One of these is -3 ms century^{-1} (tidal and other decelerating processes, such as possible $\dot{G}$ and expansion effects). The other is $+1$ ms century^{-1} (accelerating torques, possibly due to coupling effects in the Earth's core or to the draining of iron into the core from the mantle). One can therefore summarise the data examined in this section by saying that the net rate of change in the length of the Earth's day of about 2 ms century^{-1} is mainly due to tidal friction, but it *could* include a term due to a variable G with $|\dot{G}/G| \approx 10^{-10}$–$10^{-11}$ yr^{-1}, and/or expansion at a rate not greater than 0·6 mm yr^{-1}.

However, there is still no evidence definitely indicating the operation of either of the last two processes.

5.5. Secular Relaxation of the Earth's Shape

The secular acceleration of the Earth's rotation gradually slows down its spin rate, and as this process continues, the shape of the planet must alter. The secular relaxation of the shape can be defined in terms of the moments of inertia of the globe and its gravity field (the geoid) as they change through time. If part of the spin-down of the Earth is due to cosmological effects, these can in theory be calculated by studying the time evolution of the planet's shape.

Newton (1968b) has noted that data on the Devonian length of day, satellite orbits and astronomical latitude discrepancies as given by ancient eclipses, allow a possible variation in G of one part in 10^8 century^{-1}. This would lead, in turn, to an increase in the polar moment of inertia (C) of several parts per 100 since the Devonian. Most of the change in C is due to an overall expansion of the globe, this rate being compatible with the extrusion of matter taking place at the mid-oceanic ridges. Newton notes that the atmospheric tide (which contributes $+2\cdot7 \times 10^{-9}$ century^{-1} to $\dot{\omega}/\omega$) and the solar tide (which contributes $-2\cdot0$ or $-3\cdot1 \times 10^{-9}$ century^{-1} to $\dot{\omega}/\omega$), approximately cancel each other. He mentions that Dicke has used ancient eclipse data to obtain $\dot{\omega}/\omega = -15\cdot9$ ($\pm0\cdot7$) $\times 10^{-9}$ century^{-1}, part of which Dicke (see §5.6) would like to attribute to $\dot{G}$ effects. Newton considers $\dot{\omega}/\omega = -18 \times 10^{-9}$ century^{-1} to be a more reliable figure obtained from eclipse records, and is not convinced that any contribution to $\dot{\omega}$ from a $\dot{G}$ term is required. If G changes, then it produces an extra term $|\dot{\omega}/\omega| = |2\dot{G}/G|$. The larger $|\dot{G}|$ is considered to be, the more is the consequent change in C from a correspondingly smaller ancient value. In fact, Newton (1968b, p3769) is able to show that $C/M_E R_E^2$ has probably not changed a great deal in the past as a result of changes in ω, G, ρ_E or e (M_E is mass of the Earth, ρ_E is its mean density, and e is its ellipticity). If material is being extruded at the ridges and not re-subducted, then the volume of the Earth is increasing at a rate $\dot{V}_E/V_E = 5 \cdot 5 \times 10^{-9}$ century^{-1}. If C changes slowly due to this, then $\dot{C}/C = 3 \cdot 7 \times 10^{-9}$ century^{-1} leads to an evaluation of $C(\text{Devonian})/C(\text{now}) = 0 \cdot 987$. (A similar calculation by Dicke is considered in §5.6.) The concomitant expansion of the planet would have increased the temperature of the interior by about 1000 K since the Devonian. The conclusion arrived at by Newton is that C may indeed be increasing with a probability of $6:1$, but that the rate of increase is extremely uncertain. If the $\dot{C}$ term is due to $\dot{G} \neq 0$, then $\dot{G} \neq 0$ is upheld also with a probability of $6:1$. This does not necessarily mean that $\dot{G}$ *is* nonzero, since there could be independent grounds for thinking that the Earth might be expanding.

One of the main consequences of a slowing down in the Earth's rotation rate is that the excess equatorial bulge is probably in a state of slow relaxation. This will show up mainly in a change in the geoid coefficient J_2. (The geoid actually has a slight pear shape, the coefficients J_3, J_5 and J_7 being antisymmetrical about the equator, whereas J_2, J_4 and J_6 are symmetrical. The geoid shape has been refined by King-Hele and Cook (1973) to ± 1 m accuracy.) This relaxation will have serious tectonic consequences, but only if the bulge is the result of an anelastic response of the mantle to the secular deceleration of the spin. This is in turn closely connected with the viscosity (ν_m) of the lower mantle. As pointed out by McKenzie (1966), the energy stored in the excess equatorial bulge is 2×10^{30} erg (2×10^{23} J), gravitational, plus 2×10^{29} erg (2×10^{22} J), elastic. Ignoring the latter, the energy in the J_2 harmonic is seen to be greater than that in any other harmonic, which strongly suggests that it is due to an anelastic response to the spin-down with an equivalent lower-mantle viscosity of $\nu_m \approx 10^{26}$ cm^2 s^{-1} (10^{22} m^2 s^{-1}). This bulge is not due to convection currents (McKenzie 1966, p3998); its presence produces maximum shear stresses of $\approx 10^8$ dyn cm^{-2} (10^7 N m^{-2}) in the mantle. As the Earth undergoes spin-down, the moments of inertia A, A and C (C is the polar axis moment of inertia) change secularly. This long-term effect probably does not contradict those geochronological data that support a constant I (the mean moment of inertia) since the Devonian.

Differentiation of McCullagh's formula (McKenzie 1966) gives

$$\frac{dC}{dt} - \frac{dA}{dt} = M_E R_E^2 \frac{dJ_2}{dt}. \qquad (5.2)$$

If the mean moment of inertia is a constant as a function of time, then

$$\frac{dC}{dt} + \frac{2dA}{dt} = 0. \qquad (5.3)$$

Since $J_2 \propto \omega^2$,

$$\frac{2\dot{\omega}}{\omega} = \frac{1}{J_2} \frac{dJ_2}{dt} \qquad (5.4)$$

where $\dot{\omega}$ is the angular acceleration due to all causes. If $\dot{\omega}_i$ is that part of $\dot{\omega}$ due to processes interior to the Earth, then constancy of angular momentum implies

$$\frac{1}{C} \frac{dC}{dt} + \frac{\dot{\omega}_i}{\omega} = 0. \qquad (5.5)$$

Combining equation (5.5) with the previous equations gives

$$\dot{\omega}_i = -\left(\frac{4J_2 M_E R_E^2}{3C} \right) \dot{\omega}. \qquad (5.6)$$

For the Earth, the polar moment of inertia is $C \simeq 0.33 \, M_E R_E^2$, so that

$$\dot{\omega}_i \simeq -4J_2 \dot{\omega}; \qquad \frac{1}{C} \frac{dC}{dt} \simeq 4J_2 \frac{\dot{\omega}}{\omega}. \qquad (5.7)$$

Data from geochronology give values for $\dot{\omega}$, from which $\dot{\omega}_i$ and dC/dt can be extracted by the last two equations. Changes in C since the Devonian are not all that reliable, since the error in $\dot{\omega}$ is quite large.

McKenzie (1966) correctly pointed out that the rate of possible polar wandering in the Earth is governed by the region of highest viscosity, which is, in turn, inferred from data on the excess bulge. This completely upsets Gold's much quoted 'beetle-on-a-sphere' argument for polar wandering, which was based on a homogeneous Earth and led to a 10^6 yr time scale for polar wandering as inferred from the damping of the Chandler wobble (Gold 1955; polar wandering is considered at length in §5.7). The high viscosity of $\approx 10^{26}$ cm^2 s^{-1} (10^{22} m^2 s^{-1}) effectively rules out polar wandering in the Earth if the bulge is indeed of a spin-down origin. (It had been earlier suggested that it could be an effect due to the last ice age. If no relaxation has occurred since that time, a bulge of the right shape and size would have been produced, but the actual relaxation that has taken place makes this effect too small by a factor of ten to account for the observed bulge.) It is worth pointing out that convection in the upper mantle is not much affected by the high viscosity of the lower mantle. Convection might still operate in the upper region, even if phase-change boundaries are present, provided that the latter are spread out over a thickness of several hundred kilometres.

MacDonald (1964) has also considered the question of the relaxation of the bulge, and gives an interpretation of the coefficients J_n in terms of surfaces of equal density. The deviation from hydrostatic equilibrium is largest in J_2 (Mac-Donald 1964, p220), leading to the evaluation of a viscosity of the mantle (noted above) that is $\approx 10^4$ times that of the upper mantle as calculated from the uplift of Lake Bonneville. The excess bulge results in stresses in the mantle of the order of 2×10^7 dyn cm^{-2} (2×10^6 N m^{-2}) according to MacDonald, and this figure is similar to that derived by McKenzie. The energy stored in the crust due to J_2 is about 10^{29} erg (10^{22} J), and MacDonald (1964, p242) derives the total energy associated with the bulge as about 2×10^{34} erg (2×10^{27} J) which is considerably larger than McKenzie's value. If MacDonald's estimate is correct, the release of the J_2 associated energy over a period of 10^7 yr (the characteristic relaxation time) would be 2×10^{27} erg yr^{-1} (2×10^{20} J yr^{-1}); this is 0·2 of the present geothermal flux and much larger than the 10^{25} erg yr^{-1} (10^{18} J yr^{-1}) released as seismic energy. The rate of release of kinetic energy of rotation as heat, following tidal friction, is about 8×10^{26} erg yr^{-1} (8×10^{19} J yr^{-1}). In practice, the energy content of the non-hydrostatic bulge is probably somewhere in between the estimates of McKenzie and MacDonald, but either way, it represents an important aspect of the spin-down problem.

5.6. The Delineation of $\dot{G}$ Effects

The long-term spin-down of the Earth's rotation is affected by numerous geophysical processes, such as the relaxation of the bulge, that tend to mask cosmological effects. If the latter are to be delineated, the purely Newtonian terms must be considered and evaluated. Dicke's work on the secular deceleration of the Earth has been influenced by a wish to detect a term due to $\dot{G}$, as opposed to one due primarily to purely geophysical processes and expansion. The attempt to isolate such a term from historical data (Dicke 1966) is suspect because there are irregular fluctuations in the rotation that are greater than the secular decrease and may be connected with slight changes in the tilt of the axis (Dicke 1966, p111). Guinot (1970) has examined short-period terms in Universal Time, which are abrupt changes in the spin rate of order 1 ms and of 12–30 day period, tending to change the direction of the vertical and the shape of the geoid. Averaging over a period of the order of 1000 years is necessary to remove these fluctuations, but even when this is done, there remain numerous effects other than a change in G that can alter the spin rate, namely, (i) lunar, solar and atmospheric tides; (ii) angular momentum transfer to and from the solar wind; (iii) angular momentum transfer from the solid Earth to the atmosphere and oceans; (iv) electromagnetic coupling across the CMB; (v) changes in sea level that alter I, the mean moment of inertia; (vi) tectonic processes changing I; (vii) a possible continuing growth of the core causing a change in I, and (viii) a change in I due to expansion or contraction of the Earth unconnected with G, and having a thermal or cosmological origin.

Of these possibilities, (i) is known to be of major importance, but (ii) is probably negligible. The subject of interplanetary torques has been studied by Shatzman (1966), who finds that even though the Earth can support ring currents with moments of 10^{25} G cm^3 far from its axis, the direct coupling of these to the interplanetary medium cannot be significant. This does not mean that the effect of the Sun on the Earth's rotation is also insignificant, since short-term changes in the length of the day could quite well be due to winds in the upper atmosphere (§5.7) that are affected by solar behaviour. The solar cycle can obviously affect the weather, which in turn can affect the length of the day (see Meadows 1975 for a review). It has been suggested by Gribbin (1973a, c) that tides raised on the Sun

145

by the planets are the prime movers in this cycle, and the outer planets in particular align themselves, and so conjoin forces to raise tides, with a period of 179 years, a figure which appears to be reflected in the sunspot cycle. Similarly, the inner planets have a possible tide-raising potential (Wood 1972); in particular, Earth–Venus conjunctions raise tides on the Sun that are greater than the effect of Jupiter. A plot of sunspot number against tidal fluctuations caused by the four planets, Mercury Venus, Earth and Jupiter for the period 1800–1972 shows good evidence of correlation (Wood 1972), but the subject is not clear cut. In some cases, the sunspot number reaches a maximum before the peak of the tidal effects (this happened in 1950–1972), so that the influence of tides on the Sun due to the planets, reacting back on the Earth as a change in the length of the day following solar-induced weather changes, is not established beyond doubt. Winds as an intermediate agent have also been considered by Challinor (1971), but since he shows that irregular fluctuations in the spin rate for the interval 1956–1969 probably have the same source as the seasonal fluctuations, the need for invoking solar effects is not unambiguous, although a fairly convincing correlation over short terms (Olson *et al* 1975) exists between solar flares, geomagnetic storms and tropospheric vorticity patterns. The relation of phenomena like these to the Chandler wobble of the pole will be commented on in §5.7.

Continuing to examine the list of possible causes (i)–(viii) noted above, it is likely that (iii) does not have a notable influence on time scales of the order of 1000 years. Momentum transfer involving the oceans is negligible, since the angular momentum stored in the ocean currents is 10^{32} g cm^2 s^{-1} (10^{25} m^2 kg s^{-1}) or only 10^{-9} of the Earth's rotational angular momentum (Dicke 1966, p118). Leaving (iv) aside for the moment, it appears that (v) is not important unless there is no isostatic compensation following the changes in sea level involved, which is an unlikely condition. The effect is, in any case, less than 15% of the total secular acceleration (Dicke 1966, p128). Tectonic processes (vi) are similarly not expected to be a notable cause of changes in the secular acceleration; a rise of 2 km in one continent, compensated by isostasy to reduce the free-air gravity anomaly from 200 mg to an acceptable 2 mg, would only give a term in $|\dot{\omega}/\omega|$ of $\lesssim 0{\cdot}07 \times 10^{-11}$ yr^{-1}, compared with the eclipse-determined total value of about 15×10^{-11} yr^{-1}.

Growth of the core (vii) can seriously affect $\dot{\omega}$, but there are doubts attendant on the belief that any such accretion is still in progress (Wesson 1972a). If only part of the core formed in the proto-Earth, the heat released by the continuing accretion would have raised the temperature of the lower mantle by 1600–2000 K (Anderson *et al* 1972), which would be sufficient to cause more of the iron to melt and so augment the process. Initiation of the growth would eventually cause overturning of material in the proto-mantle with subsequent heating of the whole Earth by ≈ 3000 K (Dicke 1966, p136). Petrogenesis does not support the idea of core accretion (Dicke 1966, p132), and the iron would in any case have to be in bodies of size $\simeq 1$ km before it could move through a mantle of viscosity 10^{21} cm^2 s^{-1} (10^{17} m^2 s^{-1}) in 5×10^9 yr. Since the viscosity of the lower mantle is probably greater than this, it is seen that core accretion is likely to be of negligible influence, and since the displacement of iron in the Earth does not change the radius of the planet by very much (Dicke 1966, p131), it can also be ignored in expansion arguments. This conclusion could be altered if the core turns out to be of some composition other than iron, as suggested by some workers who favour a phase-change hypothesis, since cosmic abundance arguments might then imply that the iron was still in the mantle. The amount of iron in meteorites is about 27% (Dicke 1966, p134), and is therefore compatible with the amount of iron in the Earth's core, so that the geochemical grounds for suggesting that the core is not iron are distinctly dubious.

Furthermore, it should be noted that although the heat released in the solid Earth by lunar tides has not been significant during recent geological times (Munk 1966, p67), the dissipation in the early Earth might have been large enough to melt the mantle and precipitate all the iron into the core at a very early epoch (Kaula 1966). This would certainly have been important if the dissipation Q factor was such that $Q \lesssim 30$ (Kaula 1966, p50). The energy at present being dissipated in the Moon by this mechanism is about 10^{15} erg s^{-1} (10^8 J s^{-1}), but this is not sufficient to heat the interior significantly. The

Moon as a whole is able to support irregularities in its density of ≈ 10 times those of the Earth because it is a smaller body. There is therefore no reason for claiming that the Earth-facing bulge is not a fossilised tidal bulge and invoking convection to support it. The torque involved in the Earth–Moon inter-action varies inversely as the sixth power of the Earth–Moon distance and directly as the square of the Earth's radius (Runcorn 1966, p92), so it may well have been important when the Moon was nearer to the Earth and perhaps have contri-buted to the iron-draining mechanism.

It is not feasible to reach any positive conclusion about the accretion of iron by the core as it shows up in data on the acceleration of the planet (Munk 1966). Short-term changes in $\dot{\omega}/\omega$ swamp the tidal effect, and it is too much to hope that recent data will yield an estimate of the proportion of $\dot{\omega}/\omega$ caused by changes in I. This is, however, theoretically pos-sible, since, if N_y is the number of days in a year and N_m the number in a synodic month, it is shown by Munk (1966, p60) that

$$\frac{\dot{N}_y}{N_y} = \left(\frac{\dot{\omega}}{\omega}\right)_{\text{tidal}} - \frac{\dot{C}}{C}; \tag{5.8}$$

$$\frac{\dot{N}_m}{N_m} = \left(\frac{\dot{\omega}}{\omega}\right)_{\text{tidal}} (1-\gamma) - \gamma \frac{\dot{C}}{C}. \tag{5.9}$$

Here, the tidal part of the deceleration has been separated from the part due to a change in the polar moment of inertia (C), and γ is a constant ($\gamma \simeq 0 \cdot 66$). Observationally, $\dot{N}_y/N_y$ is almost equal to $(\dot{\omega}/\omega)_{\text{tidal}} \simeq -0 \cdot 23 \times 10^{-9}$ yr^{-1}, or $\dot{N}_y = -84 \times 10^{-9}$ yr^{-1}. $\dot{N}_m/N_m = -0 \cdot 07 \times 10^{-9}$ yr^{-1} or $\dot{N}_m = -2 \times 10^{-9}$ yr^{-1}. (The length of the year only changes by a negligible amount unless it is being affected by factors other than tidal friction.) Using eclipse and satellite data to obtain $\dot{\omega}/\omega$ leaves a non-tidal component of $(\dot{\omega}/\omega) \simeq +0 \cdot 04 \times 10^{-9}$ yr^{-1}. This non-tidal term is trying to accelerate the Earth's spin, but is very small compared with the net deceleration of size about $-0 \cdot 19 \times 10^{-9}$ yr^{-1}. It is the smallness of the non-tidal term that renders doubtful the existence of core accretion of iron and the hypothesis of consequent long-term changes in I, as was at one time advocated by Runcorn (1966).

148

Geochronological studies such as those of Wells (1966; see Munk 1966, p69) cannot differentiate between the tidal and non-tidal effects on the Earth's rotation so long as data are confined to the number of days in a year. If the Earth's orbital angular momentum is L_E and the Moon's orbital angular momentum is L_M, then $\beta \Delta L_M = \Delta L_E$, where β is the ratio of the solar tidal couple to the lunar one acting on the Earth ($\beta = 1/5 \cdot 1 - 1/3 \cdot 4$). Since L_E/L_M is now $\approx 10^6$, $L_E(\text{now})/L_E(\text{original}) \simeq 1$ and $L_M(\text{now})/L_M(\text{original}) \simeq 1 \cdot 016 \pm 0 \cdot 005$, and does not depend on G (Runcorn 1966, p89). From these data one can extract the change in I since, for example, the Devonian, independently of $\dot{G}$ terms. The result is $I(\text{original}) = (0 \cdot 994 - 0 \cdot 999) \, I(\text{now})$, and so is not of much use since the attendant error is $\pm 0 \cdot 003$. The reason for this is the one given above, and it now becomes obvious why it is so difficult to extract sufficient information from the Earth's deceleration data to give $\dot{I}$ and $\dot{G}$: the necessary data are simply not yet available with sufficient exactitude.

The remaining two effects, (iv) and (viii), are very difficult to evaluate. As far as (viii) is concerned, it appears that geothermal heating would cause an expansion and consequent change in the length of the day, both of which are negligible (Dicke 1966, p139). There is, however, an imposing amount of available evidence that points to the possibility of there having been a large increase in the Earth's radius since the formation of the planet (Chapter 6), perhaps connected with cosmology but not necessarily with $\dot{G}$ theories. There are no data available that contradict the acceptability of a continuing global expansion at a rate of $\leqslant 0 \cdot 6$ mm yr^{-1} in the radius, but the presence, or absence, of changes in the radius are very difficult to prove.

A similar situation holds for (iv). Dicke (1966, pp140–54) has examined this problem, showing that the non-dipole part of the geomagnetic field probably originates in the core and is connected with changes in the length of the day. The core as a whole may have a differential rotation, suggesting that it acts like a homopolar generator. Changes in the field strength can affect changes in the angular velocity of the core and so alter the spin of the whole planet. The question of whether changes in the main dipole are connected with variations in the Earth's secular acceleration is a doubtful one which rests largely on the assumed conductivity of the core material. A value $\sigma_c < 2 \times 10^{-5}$ emu seems likely, with a characteristic exponential decay time of 10^3–10^4 yr for most modes, but there are no firm data. The problem also rests on the possibility of relative motion between the core and mantle (Malkus 1966), which may be about 0.5 cm s^{-1} (so that $\Delta\omega/\omega = 9 \times 10^{-6}$). The core has a viscosity, inferred from dynamo theory, of $\nu_c \approx 10^{-2}$ cm^2 s^{-1} (10^{-6} m^2 s^{-1}) and a theoretical spin-up time from rest with respect to the mantle of $L_c(\nu_c\omega)^{-1/2}$, where L_c is the size of the system (that is, the radius of the core). Toomre (1966) notes that any steady lag in angle between the angular momenta of the mantle and core cannot be larger than 5×10^{-5} rad without conflicting with the rate of spin-down of the whole globe. Toomre draws particular attention to the inertial coupling that exists by virtue of the core precession that is consequent on the 26 000 year precession of the Earth as a whole.

Before accepting this interpretation it is necessary to rule out the other three possibilities for coupling across the CMB, namely, viscous, turbulent and magnetic coupling. (The last, as has been seen, may well be important on short time scales, and can contribute a term $|\dot{\omega}/\omega| \lesssim 1 \times 10^{-11}$ yr^{-1} on long time scales (Dicke 1966, p154).) Viscous coupling is implausible because the required viscosity of the core fluid (ν_c) is too high, although an irregular CMB may make this mechanism more reasonable. Turbulent coupling might be sufficiently large, since turbulence certainly exists if $\nu_c \lesssim 3 \times 10^2$ cm^2 s^{-1} (3×10^{-2} m^2 s^{-1}), but in the absence of firm data on ν_c it does not seem a very likely candidate. Magnetic coupling, Toomre concludes, could be effective if the field near the CMB is large, but requires the lower mantle to be a good conductor with the presently inferred CMB fields. Differential rotation of the core and mantle, as suggested by the westward drift of the geomagnetic field of $0.2°$ yr^{-1} (3×10^{-3} rad yr^{-1}), might be investigated by looking for variations in the drift of the field and working back to the variations in the differential velocity across the CMB (Munk 1966, p64). Such an approach could provide firm data on coupling across the CMB once the lower mantle conductivity is known accurately, but until such data are available the influence on ω of electromagnetic processes at the CMB remain equivocal. This concludes the survey of possible

150

effects (i)–(viii) that would have to be allowed for in calculations of the ancient spin rate of the Earth.

Despite the smallness noted previously of any substantial term not explicable by tidal friction, Dicke (1966) has sought to explain the small anomaly (6·8 parts in 10^{11} yr^{-1}) between his calculation of $\dot{\omega}$ by eclipses and the expected $\dot{\omega}$ arising from all the previously considered processes (i)–(viii). Since the total $\dot{\omega}$ being dealt with is about 20 parts in 10^{11} yr^{-1}, there does not seem to be much justification for the subsequent calculation of $\dot{G}/G \simeq -4 \times 10^{-11}$ yr^{-1}, for tidal friction alone is uncertain by a factor of two and can explain nearly all of the observed $\dot{\omega}$. Morrison (1972) has made an analysis similar to that of Dicke, and has found that there is a small difference (in the sense of spin-up) between the sum of the theoretical effects on $\dot{\omega}$ and the observed $\dot{\omega}$ as inferred from the secular acceleration of the Moon and eclipse records. This holds at the present time, but since abrupt changes in ω within the 200 years covered by telescopic observations swamp out the secular spin-down, it is difficult to say anything about the effective time scale of the acceleration process. The nature of the process causing the sudden changes in $\dot{\omega}$ and ω is also unknown, but might be connected with motion of the pole of rotation and CMB processes.

One concludes, therefore, that there may be evidence for a non-Newtonian contribution to the spin-down rate, but that the delineation of any such term is made very difficult by the presence of other processes that are trying to accelerate as well as decelerate the rotation of the planet. The margin of uncertainty in the data is such as to allow of the acceptability of a term possibly related to G variability at a rate of $|\dot{G}/G| \approx 10^{-10}$–$10^{-11}$ yr^{-1}, but this is not to say that there is any evidence positively in favour of such a term.

5.7. Rotation and Polar Wandering

It has been seen that the extraction of cosmological effects on the rotation from geophysical data is an involved process at best. In the latter part of this section the related problem of the motions of the Earth's pole of rotation is considered. Polar wandering can obscure the effects of geophysical pro-

151

cesses with a cosmological origin, although it is possible that polar wandering and fluctuations in the spin rate might themselves be initiated by cosmological effects, and their study can provide limits on such effects (Morrison 1975). Processes such as those connected with variable-G cosmologies are expected to act over long time scales, and any phenomena that are connected with them will be secular in nature. Thus it is necessary to realise that short-term fluctuations in the spin rate, such as might be caused by solar storms and similar mechanisms (Gribbin and Plagemann 1973, Gribbin 1973b, O'Hora and Penny 1973) might produce large changes (10 ms) in the length of the day, but are not relevant to polar wandering as it is usually understood because shifts in the instantaneous pole due to such processes are usually negligible. On the other hand, the Earth does have a natural wobble of its own which, while of small amplitude, is of interest because it may well be excited by earthquakes (Smylie and Mansinha 1971) and provides information relevant to the discussion of previous sections.

The Chandler motion of the instantaneous pole of rotation never grows to be very large, and so is conceptually different from the large-scale motions considered below. The period of about 400 days seems to have only one natural frequency (Pederson and Rochester 1972). The wobble is continually damped and continually excited; the latter perhaps by seasonal effects or seismic shocks. (Rochester has shown that geomagnetic core–mantle interaction cannot excite the wobble, although it can explain irregular variations in the length of the day, as discussed with particular reference to the wobble by Giacaglia 1972). The damping process is not known (Jeffreys 1970). It was suggested by Verhoogen (1974) that viscosity in the Earth's liquid core is responsible, but Rochester and others have pointed out that the viscosity, v_c, is too small by a factor of 10^{10} for this to be true, since the liquid core remains approximately undisturbed in orientation as the rest of the planet wobbles (see Rochester et al 1975). Viscous damping across the CMB would only be adequate if the wobble is excited in the core by a process connected with electromagnetic impulses, and not with earthquakes. The direct damping of the wobble by coupling across the CMB is inadequate by a factor of 10^4.

The excitation of the Chandler wobble of the Earth by random earthquakes has been held to be insufficient by Pines and Shaham (1973), who propose that the elastic energy ($\approx 10^{32}$ erg $= 10^{25}$ J) stored in the Earth in the form of its asphericity is gradually reduced by earthquakes which are triggered with a 7 year periodicity due to polar motion. Such non-random seismicity, with a certain preferential phase relationship to the Chandler wobble, does indeed seem able to keep the motion excited by the release of 10^{25} erg yr^{-1} (10^{18} J yr^{-1}) as earthquakes, this being very much smaller than the 6×10^{27} erg yr^{-1} (6×10^{20} J yr^{-1}) represented by the geothermal flow of energy.

The earthquake hypothesis is concluded by Chinnery and Wells (1972) to be reasonable, but unproven, on the grounds of a study of correlations between seismic shocks and disturbances in the path of the rotational pole. Earthquakes and the excitation of the Chandler wobble may both result from stresses set up in the mantle by impulses imparted to the outer layers by magnetohydrodynamic turbulence in the core coupling across the CMB. This point of view is due to Runcorn (see Hipkin 1970a), who calculated that if the differential rotation rate of the core and mantle is $0 \cdot 2°$ yr^{-1} (3×10^{-3} rad yr^{-1}), a sudden change of 20% of this would give a change in the length of the day of 1/300 s (as observed, for example, in 1897) provided there is electromagnetic coupling to the mantle (Runcorn 1968a). Viscous forces and topographical irregularities on the surface of the core may be able to exert forces comparable with those due to electromagnetic coupling changes, but there is a widespread feeling that it is the latter which are dominant in producing alterations in the planet's spin rate. Changes in the inertia tensor of the Earth due to faulting have been investigated by Rice and Chinnery (1972), with particular reference to the excitation of the Chandler wobble, but with no explicit mention of resulting changes in the spin rate.

Apart from the very short-term changes in the spin rate and long-term secular changes in the length of the day (due to changes in G or I), there are also changes in the length of the day on time scales of 5 years and 50 years (Munk and MacDonald 1960) of sizes ± 1 ms and $\pm 0 \cdot 01$ s, respectively. These are firmly established, unlike the very short-term fluctuations, but are not believed to have any connection with changes in G or I.

6

They cannot be accounted for by viscous or electromagnetic coupling across the CMB, although they probably have their origin in the core region (Jacobs 1972). By assuming that the inner solid core of the Earth rotates more slowly ($\lesssim 4 \times 10^{-6}$ rad s^{-1}) than the mantle (7×10^{-5} rad s^{-1}), and is supported freely in a liquid outer core of viscosity $\lesssim 30$ cm^2 s^{-1} (3×10^{-3} m^2 s^{-1}), it can be shown that turbulence in the outer core buffets the inner core (changing its rate of rotation slightly at every buffet), causing changes in the length of the day of the sizes quoted. On average, once every 10^6 years, the buffetings are statistically important enough to de-spin the inner core completely. The inner core, having a very sharp boundary (Jacobs 1972), topples over and gives rise to a magnetic reversal, the geomagnetic field in this model being possibly self-generated by a combined turbulent–homopolar dynamo. After this hypothesis had been developed to the extent of confirming its reasonableness, a model was proposed by Steenbeck and Helmis (1975) of magnetic reversals in a slowly rotating core. Their model consisted of a solid inner core rotating at a low rate inside a hollow sphere, motions between the two causing reversals in a homogeneous dynamo and rollings of the inner core. The rotation was, however, about an axis lying in the Earth's equatorial plane, and not perpendicular to it. The motion is resisted, but not stopped, by isotropic turbulence. While Steenbeck and Helmis did not examine the fluctuations in the spin rate connected with the core motions, the reality of the interrelationship was demonstrated by Press and Briggs (1975), who used a computer analysis based on a pattern-recognition algorithm to show that the Chandler wobble, earthquakes, rotation and geomagnetic changes are all connected.

Rotational changes on a 2 year time scale can be put into correspondence with alterations in zonal wind regimes (Lambeck and Cazenave (1973) have discussed the effects of atmospheric circulation). The 10 year time scale fluctuations are often interpreted as being due to motions in the liquid core. Press and Briggs (1975) note that Anderson (1974) suggested a causal relationship in which volcanic eruptions alter the wind regime and so affect the rotation (cf the discussion in §5.6 on related solar influences). This subject is a very complex one, and since there exists a correlation between peak tidal stresses and mean temperature in the northern hemisphere, as inferred from oxygen isotope studies of Greenland ice, meteorological effects are involved which are difficult to assess. Roosen *et al* (1976) have suggested that the Sun and Moon act via tidal stress variations in the Earth to influence the activity of volcanoes which, by causing changes in the atmospheric dust level, control radiation and so temperature. The observed 180 year climatic cycle may also be more directly influenced by a solar cycle of similar period (Kelly and Lamb 1976). In these ways, Earth stresses and meteorology are linked and can affect the rotation. In connection with volcanoes and tidal stresses, Johnston and Mauk (1972) have discussed the influence of Earth tides on the triggering of eruptions of Mount Stromboli. The last two phenomena could well be linked by a causal chain in which earthquakes excite the Chandler wobble, which is damped by the transfer of angular momentum to the fluid core. This results in compensatory changes in the Earth's spin, and disturbs the geomagnetic field in such a way that the effects diffuse through the mantle to appear several years later at the surface. This hypothesis does not exclude the possibility of explaining the 5–50 year rotational fluctuations in terms of processes to do with the core.

It is as well to remember that while changes in the spin rate of the Earth on time scales of 5 and 50 years, as discussed above, are identifiable, there is almost certainly a continuous spectrum of time scales for fluctuations in the length of the day, ranging from thousands of millions of years to periods of a month or so, and perhaps even less. The variation in rotation on time scales of 5–50 years can be physically connected with polar wandering of the solid inner core of the Earth. Fluctuations in the spin rate on longer time scales (connected with cosmological effects) could be obscured by large-scale polar wandering, to which the rest of this section is devoted.

Polar wandering, as conventionally understood, refers to motions of large parts of the Earth relative to a fixed direction in space with an amplitude ≈ 1 rad, which is much larger than the Chandler wobble. It means either (*a*) a shift in the spin axis of the Earth, or (*b*) a shift of much of the Earth with respect to a fixed rotational axis, or (*c*) a shift of the surface layers of the

Earth over those underneath (and vice versa). The latter is dynamically feasible, but should be sharply differentiated from (d) magnetic polar wandering, which is a term employed here to mean a shift in the configuration of the Earth's magnetic field with no shift in the solid globe.

Of the types (a)–(d) of polar wandering just noted, type (a) is very unlikely unless large bodies such as giant meteorites strike the Earth. This has been proposed by Gallant (1963), but seems unwarranted. Alternatively, Mateo (1972) has suggested that if the centre of the Earth's core were primaevally offset from the overall centre of mass, then the tendency to coincidence of the two centres would result in a motion of the pole at a rate of 1–10 cm yr^{-1}. There does not appear to be any notable evidence of the operation of this type of process.

The process (d) is attractive because it does not run into many of the problems, examined below, met by polar wandering of types (b) and (c). The main difficulty for (b) and (c) is that the Earth's shape, and density inhomogeneities within the planet, tend to prevent solid-Earth polar wandering. Magnetic polar wandering implies some shift in the geomagnetic poles only, with at most a reorganisation of material in the Earth's core and no overall movement of the upper layers. There is some evidence of rapid shifts of palaeomagnetic poles that may support this type of wandering (see Wesson 1972a). Palaeomagnetic polar wandering curves imply true wandering only if pole positions derived from data from different continents yield the same wandering track. If the palaeomagnetic pole tracks do not coincide for different continents, then continental drift is usually inferred. Some examples of coincident tracks are known and, in particular, the paths are coincident for data from North America and Africa for the period 2700–800×10^6 years ago in the Proterozoic (Piper 1976, Smith 1976). These data are not, however, undisputed, since McGlynn et al (1975) have interpreted $100°$ latitude shifts in the period 2200–1800×10^6 years ago in terms of plate tectonic movement at continental drift rates of about 3 cm yr^{-1}. It, should be noted, though, that palaeomagnetic polar wandering tracks of more recent date are often orientated along lines of (present day) longitude, suggesting a purely geomagnetic effect. If polar wandering of type (d) is to be invoked to explain evidence indicating wandering, then a detailed mechanism of magnetic polar wandering that includes reversals is needed.

With regard to the magnetic reversal problem, several mechanisms of reversal have been suggested that are particularly pertinent to wandering of type (d). These can involve changes in the convection regime in the outer core (Jacobs and Masters 1976), or some kind of permanent magnetisation of the inner core with a strength-varying magnetohydrodynamic field of opposed polarity in the outer core (Verosub 1975; see below). These mechanisms involve shifts of the whole field. An alternative mechanism is one in which oppositely directed fields in rock samples are induced in the rocks themselves with no realignment of the main field. Ryall and Ade-Hall (1975) have caused self-reversal of thermoremanent magnetisation in pillow basalts using laboratory methods. This shows that self-reversal can occur when the local temperature has been increased after acquirement of magnetisation either in the oceanic crust or in the continental lithosphere with hydrothermal intrusions. Such temperature increases can split the cation-deficient titanomagnetite into two or more phases, leading to self-reversal. The mechanism has been further discussed by Stephenson (1976), and it could account for those periods recorded by palaeomagnetic data in which the geomagnetic field appears to have been anomalous. The Ordovician was such a period; data from this epoch indicate that the apparent time between reversals may have been 10^7 years then, as opposed to the 10^5–10^6 years of more recent times (Thomas and Briden 1976). Provided the question of the nature of palaeomagnetic reversals can be adequately answered, it would seem that magnetic polar wandering of type (d) could represent a viable process.

Solid-Earth polar wandering of types (b) and (c) is a much more problematical process despite having been more fully investigated. Most of the rest of this section will be devoted to a study of some of the problems of solid-Earth wandering. It should be realised, however, that there is no evidence available that definitely shows the effects of such wandering. This may be because geodynamical processes have since destroyed the effects of such ancient polar wandering in the case of the Earth, whereas they might have been expected to survive in a

157

less active planet. For Mars, it has been proposed by Murray and Malin (1973) that polar wandering may be occurring at a rate of about 5×10^{-9} rad yr^{-1} ($\simeq 20$ arc min per 10^6 yr). The difference in cratering density between two reference hemispheres and the structure of the Coprates assemblage on Mars have been taken by Countillet and Allegre (1975) to indicate the action of plate tectonics and by McAdoo and Burns (1975) to indicate polar wandering. There is a system of faults on Mars that could be explained on the basis of an idea, attributable to Vening Meinesz, that a planet undergoing spin-down changes its shape to become less flattened, so setting up stresses that result in a planet-wide system of lineaments. Those observed on Mars run predominantly NW and NE (Binder 1966, Binder and McCarthy 1972). Alternatively, and perhaps more likely in the instance of Mars (which is undergoing only negligible tidal spin-down), the faults could be due to a shift of the surface layers of Mars over an underlying oblate spheroid. Such a motion would set up membrane stresses (of the order of 10^{10} dyn $cm^{-2} = 10^9$ N m^{-2}) much greater than the attendant bending stresses, and this mechanism of polar wandering appears to be a viable one.

Polar wandering of types (b) or (c) has been implicated by Williams, who has commented on the existence of broad-leaved plants at high palaeolatitude during the Tertiary (Williams 1974). Such plants would have needed a light flux most readily provided by a globe tilted at a different angle to the ecliptic, if they were similar to modern plants of the same type. Further evidence of polar wandering comes from Hargraves and Duncan (1973), who have suggested that the mantle of the Earth rolls independently of the lithosphere, with the crust fixed relative to the rotation axis (a combination of types (b) and (c) above). This deduction rests on an analysis of eight mantle plumes, which show location discrepancies between palaeomagnetic and volcanic trace data. These authors do not examine the effects such a motion would have on the Earth's core and the length of the day. It may be that the data are better explained by invoking movement of the plumes only with respect to each other, although, as noted above, rolling of the *inner core* is compatible with the known fluctuations in the spin rate.

It seems highly unlikely that the mantle or any large part of the Earth, except possibly the inner core, could shift considerably. This is because the high viscosity of the mantle (§5.5) and the density inhomogeneities present in it tend to prevent any polar wandering. There is good evidence for lateral inhomogeneity in the uppermost 200–300 km of the mantle (Hales and Herrin 1972), probably due to temperature effects (Herrin 1972). The non-dipole magnetic field of the Pacific has long been recognised as unduly subdued, and this may be connected with lateral inhomogeneities in the lower mantle, perhaps coupled to the core (Doell and Cox 1972). The fact that the westward drift of the non-dipole part of the field seems to be centred about an axis not coincident with the geographical one (Malin and Saunders 1973) should not be taken as evidence in favour of polar wandering (Tanguy and Wilson 1973), although it does tend to support the hypothesis that the pole of rotation of the outer core is slightly different from that of the mantle. A maximum in the drift rate, which proceeds at about $0.2°$ yr^{-1}, occurs when the rotation pole is near the geographical pole, suggesting that the offset of the two axes may not be permanent. The geomagnetic and seismic evidence showing that density inhomogeneities exist in the mantle, and also in the crust (Woollard 1972), means that the rotation axis cannot move with respect to the inertial axis except over time scales long enough to alter the configuration of the mantle inhomogeneities.

Convection currents could, of course, be responsible for, and alter, the inhomogeneity configuration in time, leading to large-scale polar wandering (Takeuchi and Sugi 1972), but there are problems with this hypothesis. Firstly, the upper-mantle inhomogeneities, as discussed in the last paragraph, cannot produce unlimited polar wandering of type (b) because the rate is fixed (virtually at zero) by the zone of highest viscosity which, as noted previously, is the lower mantle. Convection, especially of irregular type, can still occur in the upper mantle and manifest itself as sea-floor spreading, where, as indicated by Pollack (1969) and Lister (1975), it can provide a gravitational drive for plate motions. Such models can be described by boundary layer theory, which is based on the gravitational sliding, and contraction by cooling, of plates moving away from ridges and floating on an asthenosphere of viscosity $\approx 10^{20}$ cm^2 s^{-1} (10^{16} m^2 s^{-1}) that is ≈ 100 km thick. Such models, which are similar to that of Wesson (1972b), are even more attractive if linked with expansion as the originating mechanism for the ridges (§6.2), conceivably coupled with phase changes. It must be admitted that upper-mantle convection may be compatible with polar wandering of type (c) caused by cell reorganisation and slippage over the upper/lower mantle interface. Secondly, however, one must note the empirical fact that global heat flow is low over Pre-Cambrian shield areas and the oldest parts of the ocean floor (Chapman and Pollack 1975). This suggests that the thermal regime and sial–sima heat flow topography of the upper mantle are stable phenomena (cf §6.5) not disturbed by solid-Earth polar wandering, although consistent with fixed convection patterns. Thirdly, the upper-mantle inhomogeneities, even if dynamic entities, are supplemented by more fundamental anisotropies. It has been shown that inhomogeneities in seismic velocities delineate ocean–continent demarcations down to 400 km, and that lateral velocity gradients extend to a depth of $\gtrsim 800$ km (Jordan 1975). These could, it is true, be interpreted as being due to deep convection, but this concept is unlikely (Wesson 1972a, b). Among other things, it is in contradiction with the oxidation state (ferric/ferrous ratio) and nickel content of rocks at the Earth's surface as compared with the equilibrium chemistry characteristic of the Earth's core.

For the three reasons given, it seems very likely that the presence of inhomogeneities not attributable to convection currents within the Earth must severely restrict large-scale motions of the whole planet. The realisation of the existence of these inhomogeneities has made obsolete many claims that tectonic processes can alter the moment of inertia of the globe in different regions and so instigate polar wandering of types (b) or (c) above. This criticism applies to the work of Inglis (1957), who has claimed that the pole is stabilised at its present position by the melting of circumpolar ice, and that the excess equatorial bulge can adjust when the pole moves away from this position. The latter claim is without basis unless some origin of the bulge is postulated other than that of anelastic spin-down. The argument by Gold (see Inglis 1957, p17) that

160

a rise in level of a continent by 30 m in 10^5 yr can cause polar wandering of the order of 1 rad in that time is not applicable to the Earth as it is now understood, since the argument depends on the assumption of a spherical, homogeneous globe. (This should not be confused with changes in the lunar obliquity, and possible small shifts in the Earth's axis due to tides (Gold 1966), these being very long-term effects.) Likewise, it used to be thought that palaeoclimatic evidence in favour of continental drift could be interpreted in terms of polar wandering, but a review of the palaeoclimatic data by Stehli (1973) leads him to the conclusion that no results which are unequivocal in interpretation exist at present. Other aspects of this problem are discussed by Wesson (1973). Although continental drift and/or polar wandering are widely accepted today, even a casual perusal of results on the relative polar wandering tracks of Africa and North America (Spall 1972), collisions of plates (Roman 1973), and the movements of Africa (Pawley and Abrahamsen 1973) and India (Blow and Hamilton 1975), serves to show that the evidence for drift or wandering is inconclusive. Gilluly has given a list of observations connected with mountain ranges that do not fit into the conventional picture of plate tectonics (Gilluly 1972), while Kaula (1972) has drawn attention to the anomalous position of Antarctica with respect to the convection-interpreted geoid. When plate motion and polar wandering occur together (if they ever do), the usefulness of polar wandering can only be decided when the relative motion between all the major plates and the pole has been determined (McKenzie 1972). This cannot be done until much more data are available.

A strong argument against polar wandering can be made by considering the excess equatorial bulge (§5.5). Any shift of the axis of rotation is opposed by the bulge, just as it is difficult, for example, to deflect or tilt a rapidly spinning bicycle wheel held between the hands. At the present time there is no conclusive argument against the hypothesis that the bulge is a result of imperfect relaxation of an anelastic lower mantle with an effective viscosity $\nu_m \approx 10^{26}$ cm^2 s^{-1} (10^{22} m^2 s^{-1}), although arguments against this interpretation have been made, especially by Dicke (1966) and Goldreich and Toomre (1969). Dicke's argument is that the e-folding decay time of a spherical harmonic distortion is proportional to $p\sqrt{p+1}$, where p is the degree of the spherical harmonic (Dicke 1966, p122). Dicke (1969) has discussed the response of the asthenosphere during post-glacial adjustment. Fennoscandian data ($p=25$) give a time of relaxation of the order of thousands of years, while Lake Bonneville ($p=100$) has a relaxation time of about 4000 years. If, therefore, the $p=2$ harmonic had the same type of origin as the higher-degree harmonics, the time constant would be only 1000 years. (This is for a model of the Earth composed of a uniform Newtonian fluid of viscosity 10^{21} cm^2 s^{-1} (10^{17} m^2 s^{-1}).) Dicke is thus forced to conclude that the bulge is supported by whole-mantle convection (Dicke 1966, p126). He bolsters his contention by pointing out the nonexistence of 100 mg gravity anomalies that would be present over large portions of the crust if the model used were incorrect. The fault in Dicke's approach is that it is not certain by any means that the $p=2$ harmonic does have the same origin as the higher-degree harmonics, and evidence to the contrary has been provided by McKenzie (1966). Further, there are gravity anomalies of even Palaeozoic age that have survived to the present day despite the prediction of Dicke's simple model that they should have subsided long ago (Wesson 1972a). It is thus doubtful whether Dicke's argument is plausible.

An alternative hypothesis for the origin of the excess bulge has been suggested by Goldreich and Toomre (1969). They showed that if the bulge is excluded from account, the figure of the Earth is triaxial and the present rotation axis is not preferentially aligned with the axis of greatest moment of inertia. They therefore infer that polar wandering might take place at a rate of about 10 cm yr^{-1}, depending on the viscosity of the mantle (which their analysis puts at $\nu_{\mathrm{m}} \lesssim 10^{25}$ cm^2 s^{-1}, and probably an order of magnitude smaller). The work of Goldreich and Toomre is considered by Cathles (1975) as demonstrating that the excess (non-hydrostatic) part of the bulge is in fact only an artefact of the spherical harmonic analysis used to express the gravity field, and Cathles therefore believes that the excess bulge no longer provides evidence that the lower mantle has a high viscosity, preferring instead to believe that $\nu_{\mathrm{m}} \lesssim 10^{22}$ cm^2 s^{-1} (10^{18} m^2 s^{-1}). The argument of Goldreich and Toomre has been criticised by McKenzie (1972, p355), and the low-viscosity mantle which the hypothesis would imply seems to be in conflict with other data, to do with Earth–Moon tides and the decay of gravitational anomalies on the Earth, which favour a high-viscosity lower mantle (Jeffreys 1970). If the excess bulge were indeed of negligible importance, the rotational axis of the Earth and the axis of maximum moment of inertia would tend to line up (Goldreich and Toomre 1969, Pan 1972), producing rapid polar wandering. There is evidence from palaeomagnetism both for and against wandering at the rate noted above (Wesson 1972a, McKenzie 1972, p356). On the other hand, the basic point of the Earth having a triaxial, non-aligned figure once the bulge is taken out of account would seem to be correct, and some small secular motion of the pole is possible in principle, depending on how strongly it is resisted by other factors. The question of polar wandering thus depends on whether it is possible for the Earth to deform in a way such as to eliminate the deviation between the rotation axis and the axis of maximum moment of inertia. If so, then the geoidal axis will drift towards the rotation axis with the axis of figure being dragged along, as discussed in an idealistic case by Gold (1955). The factors which resist the motion are (i) the high viscosity of the lower mantle, whether deduced from the existence of the bulge or from one of the other pieces of evidence considered by Jeffreys (1970) and Wesson (1972a); (ii) the bulge itself, and (iii) any other mass inhomogeneities within the Earth. Although one should keep in mind the implications of the analysis by Goldreich and Toomre, the effects of (i), (ii) and (iii) must severely restrict the admissibility of extensive polar wandering. Large movements ($\approx \pi$ rad) of the rotational axis with respect to the Earth's surface are therefore unproven, and there are cogent reasons for believing that they cannot occur.

5.8. Results on $\dot{G}$ from the Rotation

It has been seen that when the secular deceleration of the Earth's rotation is isolated from the various types of fluctuations affecting the spin rate (§5.2), the secular acceleration of the Moon in its orbit can be obtained. Provided the tidal part of the acceleration $\dot{n}/n$ can be removed, $\dot{G}/G$ can be calculated

163

(§5.3). The removal of the non-cosmological terms is a notorious problem that depends on the correct estimation of contributions due to tides (both oceanic and in the solid body of the Earth), and to global expansion and geophysical effects that give rise to discrepancies between the spin-down rate as inferred from palaeontological data (§5.4) and more modern methods. The length of the day has been increasing at about $2 \text{ s } 10^5 \text{ yr}^{-1}$, and this causes a secular change in the Earth's shape (§5.5) that needs to be taken into account before the part of the deceleration due to $\dot{G}$ (or other cosmological causes) can be isolated. When all the reasonable geophysical terms have been considered (§5.6), the end result is *consistent* with a finite value for $\dot{G}$ of $\dot{G}/G \simeq -4 \times 10^{-11} \text{ yr}^{-1}$. It is also consistent with other cosmological processes, such as an expansion of the Earth, that give an equivalent dynamical behaviour (that is, processes such as expansion (Chapter 6) with G constant that can be parametrised in a variable-G formalism, if desired). It should be strongly emphasised, however, that while studies of the rotation of the Earth are compatible with $\dot{G}/G \simeq -4 \times 10^{-11} \text{ yr}^{-1}$, or the expansion hypothesis with a secular increase in the Earth's radius at an equivalent rate given by equation (5.1), this does not constitute a proof of such effects. The reason lies in the lack of exact data and in the difficulty of estimating, or finding sufficient grounds for ignoring, large-scale geophysical processes such as polar wandering (§5.7). Despite this ambiguous situation, certain methods can be employed to short-circuit some of the geophysical effects and arrive at definite results of cosmological interest.

An explicit attempt to use the rotation of the Earth to estimate the possible rate of change of G has been made by Morrison (1973), who has examined timings of 40 000 lunar occultations of stars over the period 1943–1972 and confirmed that the length of the day has been steadily increasing but with superimposed fluctuations of short duration. The rate of increase from recent observations (4 ms century^{-1}) is somewhat greater than the 2 ms century^{-1} given by ancient eclipse data and the 1·5 ms century^{-1} given by averaged telescopic data over the period 1663–1972. Attempts to account for the difference as an effect of $\dot{G} \neq 0$ must be judged carefully since the Moon's motion is governed by a gravitational time scale, whereas

Morrison's data employ Atomic Time. Elsmore (1973) has discussed ways of measuring time that are relevant to $\dot{G}$ experiments, while Martin and Van Flandern (1970) have considered the effects of tidal friction in altering the elements and shape of the lunar orbit. Morrison is able to conclude that $|\dot{G}/G| \lesssim 2 \times 10^{-11}$ yr^{-1}. The fluctuations in the rotation rate, which seem to be real, still need further explanation.

Work such as that of Morrison (1973, 1975) on the spin-down suggests that one might obtain limits on $\dot{G}$ by looking for indirect effects of the deceleration of the Earth on its tectonic features. This might also be a way to detect effects of a possible global expansion. A train of reasoning one can employ is as follows. (i) the Earth is undergoing spin-down, leading to (ii) a secular change in the ellipticity which shows up as (iii) a secular change in J_2. This causes (iv) a secular change in the gravitational potential which (v) should produce a pattern of lines of compression and tension leading to the expectation (vi) of finding a regular system of fold mountains and rifts, with enhancement of the geothermal flux and seismic energy release near mid-latitudes (45°). The expected features (vii) are not seen, and this null observation suggests either that G has changed or that some great changes may have taken place in the radius. One can infer from this line of reasoning that the absence of a regular system of mountains and rifts implies either that G has indeed been varying at a finite rate of $|\dot{G}/G| \gtrsim 10^{-12}$–10^{-13} yr^{-1}, or else that the planet has experienced a large change in radius since Pre-Cambrian times.

Van Flandern (1975), using a set of data similar to those of Morrison, carried out an important analysis in which he determined the secular acceleration of the Moon's mean longitude to be $-65''$ (± 18) century^{-2}, on Atomic Time. The tidal component of $-38''$ (± 4) century^{-2}, which is measured using Ephemeris Time, can be removed, leaving $-27''$ (± 18) century^{-2}. This was interpreted by Van Flandern as resulting from a finite rate of change in G at a rate $\dot{G}/G = -8$ (± 5) $\times 10^{-11}$ yr^{-1}, although he clearly realised that it is also consistent with the expansion hypothesis. There has been some discussion of the meaning of this result, and in particular, the doubt has been raised that his occultation observations cover a period of time (1955–1974) such as might allow a serious source of systematic

165

error to creep in connected with the 18·6 year period of the motion in longitude of the Moon's ascending node. This possibility was, however, considered by Van Flandern (1975, p342), and was incorporated in an increase of the quoted error from ± 3 to the noted $\pm 5 \times 10^{-11}$ yr^{-1}. This error is an upper limit, so that the result is formally significant. The two alternative explanations—expansion and mass loss—which are consistent with Van Flandern's calculation will be considered in detail in Chapter 7 from a theoretical point of view. It will be seen that expansion is the most reasonable alternative to a finite rate of change in G. Jordan (1971, p114) interpreted the increase of the length of the day on Ephemeris Time as due to expansion of the Earth at a rate of 0·52 mm yr^{-1} with no change in G. Van Flandern's result as it stands would seem to be valid, and is clearly of great significance whether taken on the $\dot{G}$ or expansion basis, the latter being especially important for geophysics.

6. *The Expanding Earth*

6.1. Introduction

The hypothesis that the Earth could be in a state of secular expansion has a history that is almost as long as that of continental drift. It was in a rather inconsequential condition prior to the appearance of the modern cosmologies described in Chapter 2, because there was no theoretical basis on which to justify the hypothesis. McVittie (1969) has given an excellent review of the basic implications of cosmologies in which $\dot{G}$ is finite, with particular reference to the expanding Earth hypothesis. The variable-G theories give adequate reason for considering the evidence which tends to show that the Earth is expanding, and there are several other well founded hypotheses of a fundamental nature that would have similar consequences.

6.2. Evidence in Favour of Expansion

When engaged in continental drift reconstructions, Creer (1965a) made the interesting discovery that all the continents fit closely together on a globe about half its present size. It is now realised that expansion from a completely sial-covered globe of about 3700 km radius, at a constant rate of about 0·6 mm yr^{-1} over $4·5 \times 10^9$ yr, would give continents a configuration as we now see them, perhaps modified by continental drift (Wesson 1973, 1974b, c). The expansion would involve distortion of the continents by about 20%. This was discussed by Jeffreys (1962), Dennis (1962) and Barnett (1962), with respect to the complete-crust reconstruction of the latter worker.

Dearnley (1965, 1969) made an exhaustive study of expansion in relation to orogenic fold belts and convection, and came to the conclusion that data from the Superior, Hudsonian and

Grenville regimes dating back 3.5×10^9 yr all fall on a line which implies expansion at a steady rate of 0.65 (± 0.15) mm yr^{-1}. The implication of convection currents in this work is rather unfortunate, and, it would seem, unnecessary (Wesson 1971, 1974b, c), because they are based on the old idea of accretion of iron by the core controlling the order of the convection pattern in the mantle. This idea contradicts several geophysical data (Wesson 1972a), and is probably in error, as is the belief that Eötvös forces have any noticeable influence on the dynamics of the Earth's interior (Jeffreys 1970, p451). Despite these defects, Dearnley's work stands as interesting evidence in favour of the expanding Earth hypothesis, as does Barnett's work on oceanic rises (Barnett 1969), and Meservey's demonstration that continental drift is topologically inconsistent if reconstructions are made on the present-sized globe (Meservey 1969).

The tectonic necessity for an expanding Earth was only realised some time after the hypothesis first started to gain support. The original need for it was pointed out by Egyed (1956) on the basis of palaeogeographical maps by Termier and Termier and Strahov. These showed a gradual increase in the amount of land above sea level over geological time, although some fluctuations were superimposed on this trend. This implies either that the oceans are decreasing in volume (which may not be acceptable, since the biosphere has actually grown by about 4% since the Cambrian), or else that the Earth is expanding. The palaeogeographic evidence has been re-emphasised by Termier and Termier (1969) in a most comprehensive account covering the whole history of the Earth from Eocambrian to Recent, although the expanding Earth interpretation has been criticised by Armstrong (1969), who would prefer to explain the data by a combination of processes involving polar and glacial ice, mountain building, erosion and oceanic rises. These influences can indeed explain the observed fluctuations in the data as transgressions and regressions on a time scale of 10^7–10^8 yr, but it is not clear whether they can explain the general trend towards continental emergence. More volumetric figures are needed than those given by Armstrong. Veizer (1971) has also claimed that palaeogeographical maps give a wrong impression that there were more water-covered

168

areas in the past. The chance of a transgression or regression over a particular continental block increases as the time interval considered increases, old rocks having suffered more epierogenic, orogenic and isostatic movements in comparison with recent rocks, the latter having fuller geological sequences. This is connected with the association of low sea level with ice ages and of high sea level with carbonate deposition and transgressions. The main point of the criticism is undoubtedly valid, but it is necessary to show that this selection effect has indeed been present in the work of Strahov, the Termiers and Egyed, and that, if so, it is sufficient to account for the large trend in their data without appeal to additional hypotheses. Continental accretion falls into the latter class, but is usually introduced in an *ad hoc* and implausible way as an adjunct of convection.

Hallam (1971) has re-evaluated the palaeogeographic argument, preferring instead an explanation by thickening of continents from below. This new hypothesis, however, incorporates an *ad hoc* eustatic change in sea level, and it is not obvious that a simple increase in crustal thickness of 600 m would give the growth needed of 1 mm per 10^3 yr. Loading problems of this type are notoriously difficult (Jeffreys 1970, pp417–24), and have long been a geological puzzle, for example, when applied to geosynclines. On the whole, it seems that expansion is still the best explanation of continental emergence, and Egyed has restated reasons for believing it to be valid (Egyed 1969). He has pointed out that the Moho frequency–depth curve and the hypsometric curve find natural explanations as consequences of the breaking up of a sial cover under the force of expansion at a rate of 0·6–0·7 mm yr^{-1}. Egyed attained the quoted rate of expansion by adding together terms due to Dirac's cosmology, a Ramsey high-pressure phase change (Urey 1952, p69) and heat expansion caused by the natural radioactivity of the Earth.

It is often stated that the Ramsey phase-change hypothesis for the core–mantle boundary (CMB) is unacceptable, but it is, perhaps, only fair to refer to some work on the problem by Lyttleton (1965), whose model involves substantial changes in the Earth's radius. The proto-planet is envisaged as liquefying at the centre, causing a catastrophic collapse in which about

6% of the entire mass liquefies and the radius decreases by 70 km, this being followed by a slow contraction of the whole (now stable) planet. Lyttleton thus favours contraction of the Earth, with periods of slow expansion in between, where an overall decrease in radius (by 370 km) is inevitable. It does not seem, however, that the theory is applicable to the Earth as we now know it, since the contraction leads to a decrease in the length of the day (Lyttleton 1965, p486), which goes against fossil data (§5.4). Little information is given about the mechanism of the phase change on which the model relies, and Lyttleton, in fact, seems to confuse phase changes with electron degeneracy due to high pressures, the latter being a phenomenon requiring pressures much greater than those obtaining at the centre of the Earth (Holmes 1965). The ionisation imagined by Lyttleton (1965, p487) plays a part in explaining the geomagnetic field as a result of the usual dynamo action, but with the generation proceeding in a non-iron core which is rotating faster than the mantle. The core model is unlikely, since the abundance of iron in the Earth is now known to be not anomalously high compared with the solar abundance, so there is no good reason for denying that the Earth's core is, after all, mainly iron. Frank (1971) has criticised Lyttleton's extension of Ramsey's hypothesis on thermodynamic grounds, but his work is in error. Lyttleton has applied his theory to the other terrestrial planets (Lyttleton 1970), assuming that they all originated from a primaeval material of the same composition. The approach works satisfactorily for the Moon, Venus and Mars, but as is usual with theories of the composition of the planets, Mercury does not fit into the scheme. In view of these objections, and the fact that palaeogeographical and other evidence points to expansion and not contraction of the Earth, Lyttleton's hypothesis as it stands is not much supported by available evidence.

Lyttleton (1970) explicitly refuses to consider the possibility that G is variable, since he considers this to be a logical mistake. Nevertheless, his work has been valuable in reinforcing the demonstration that whole-mantle convection is impossible as a means of supporting the Earth's much discussed excess hydrostatic bulge (Lyttleton 1970, p117; §5.7), since that would need flow velocities of 10^2 cm s^{-1}. It is also useful in demonstrating that Runcorn's theory (Runcorn 1965), in which changing convection modes control the development of continental drift, is mathematically questionable (Lyttleton 1972). The polytropic solutions of the Earth's structure, as determined by Lyttleton, are used by Hoyle and Narlikar in evaluating the geophysical consequences of their cosmology (§2.8). This cosmology does lead to expansion, and so does not conflict with the palaeogeographical data, although maps incorporating such data are perhaps best considered only as qualitative support for the expanding Earth hypothesis, in view of the objections to the palaeogeographic method noted above.

Expansion of the Earth at a rate of about 0.6 mm yr^{-1} can be employed in geophysics in several ways. The mid-oceanic ridge system, for example, surrounds Antarctica almost entirely, yet convection current theory in its presently accepted form must then imply that the continent is shrinking or expanding. Expansion of the Earth is a more likely alternative (Wesson 1972a). This would also help in the cases of other ridges, in the processes of orogenesis, and in explaining how the continents are maintained above sea level (Wilson 1961). It is difficult to escape from the dilemmas posed in the situations of Antarctica and Africa by postulating that just these two continents are changing in size. Besides being hopelessly *ad hoc*, this would lead to distortion, as shown by Jeffreys (1962), and so would invalidate once again the basic premise of continental drift and plate tectonics, namely, that the continents have not changed shape by large amounts, at least since the Pre-Cambrian.

That the mid-oceanic ridges look like cracks in an expanding globe is obvious (Heezen 1960), and it is interesting to see if the area of the ridge system can be accounted for by expansion. This was investigated by Wilson (1960), who adopted the decreasing-G hypothesis in accounting for the observed ridge system area of order 10^8 km^2. This is conformable with the expansion hypothesis when it is recalled that the existence of the ridges since the expansion began, and their length with respect to the present $\approx 10^5$ km, are both questionable subjects. The possibility that the ridge system has not always been so extensively developed is suggested by some calculations of Tarling (1975), who believes that the volatiles and water released by sea-floor spreading along the ridges at reasonable rates inferred from plate tectonics might have resulted in the loss of 10^{24} g of mantle material during the last 2×10^8 yr. This is a very high turnover rate, and would imply that the present oceans of the world have an age of only $\approx 10^8$ yr. If true, it would enhance the expanding Earth argument given above, but is questionable because it is in disagreement with the more moderate rate of biosphere growth noted previously.

Approximately 85% of the world's ridge system lies on circles around continental shields, while 50% of the total is centred in ocean basins (Menard 1965). It is important to realise that the ridges, although bounded by 200–300 km wide regions of high heat flow, are not exceptional seats of magma outpourings, a picture often depicted by proponents of whole-mantle convection. The volume of extruded basalt on land is of the same order as that in the ocean basins (Menard 1965, p113), when averaged over long periods of the Earth's history. These facts point to a phenomenon affecting the whole globe. In particular, the Darwin rise, which is composed of the order of 10^7 km^3 of basalt (as is the mid-Atlantic ridge) shows evidence of an extensional nature conformal with the expanding Earth hypothesis (Menard 1965, p118). It has been suggested that the tectonic fabric of the sea floor is better explained by expansion than by convection currents in the mantle (Heezen and Tharp 1965), and it has even been suggested that islands in the Indian Ocean, the Alps, and continental drift all owe their existence to expansion. The energy budget of the Earth is expected to undergo some modification if it drives such orogenic processes by expansion. It was to prevent the whole globe from melting that Urey (1952, p176) was obliged to restrict the amount of expansion to a few hundred kilometres on his core compression hypothesis. Urey was, of course, in favour of a cold origin for the Earth. If it is desired to have a molten Earth, then expansion by 500 km during reorganisation of the proto-planet can easily cause the melting, and the greatest release of energy would be near the CMB, as shown by Beck (1969). Urey and Beck do not consider a cosmological contribution to expansion. Jordan (1969), however, on the basis of his $G \propto t^{-1}$ theory, would prefer to

172

explain the observed process of sea-floor spreading in terms of expansion of the globe.

If one imagines a great circle drawn around the Earth and an average sea-floor spreading rate from ridges of a few centimetres per year, it is seen that a rate of increase in circumference of the globe of the order of 10 cm yr^{-1} is needed. It has been suggested by Steiner (1975) that the rate of opening of the Atlantic may need complete sea-floor spreading rates of up to 12·5 cm yr^{-1}, although this depends on the interpretation of some palaeomagnetic results that are disputed as regards significance. Lower sea-floor spreading rates may be acceptable because of palaeomagnetic ambiguities, but a value of about 10 cm yr^{-1} overall is still required. Unfortunately, the expanding Earth hypothesis, with dR_E/dt of about 0·6 mm yr^{-1}, can only give a circumferential increase of about 4 cm yr^{-1}. To get around this, one might postulate that the observed spreading rate at ridges is higher than the average due to concentration there by some mechanism. The intriguing idea quickly follows that sea-floor spreading may be something of an illusion caused by continents remaining still while the globe expands beneath them. This picture fits in with the model of the Earth discussed by Wesson (1974b, c) in which the continents are secularly stationary in a lateral sense with respect to each other, but with sea-floor spreading active in the ocean basins owing to convection in the upper 700 km of the mantle, of a very restricted type (see also §7.5). Active orogenetic leading edges and the corresponding trailing edges retain their meaning only in comparison with the mobility of the ocean floor. The common picture of convection currents rising from depth, causing large heat release behind island arcs and so on, due to adiabatic decompression, is in difficulties here (Waldbaum 1971, Ramberg 1971), since a temperature rise of only 20 K can be obtained by Joule–Kelvin decompression (Dodson 1971). A model involving shallow convection seems to be preferable. The heat released by viscous dissipation is probably the explanation of high heat flow behind island arcs. On the whole, there is no need in geophysics for deep-seated convection.

If material is to rise from the lower mantle, many of the problems attendant on conventional convection can be avoided by postulating that it ascends in pipes or plumes (§5.7). Stewart has assumed that hot spots in the lithosphere are due to plumes that are anchored in the mantle, and has analysed evidence for the lateral motion of such hot spots in an attempt to measure the rate of expansion of the Earth (Stewart 1976). Using data for hot spots of three ages (the present, 5×10^6 years ago and 120×10^6 years ago), he has investigated the great-circle distance between members of pairs of spots. While the errors are large, his measurements indicate that typical hot spots in the lithosphere are moving with respect to each other at relative speeds of $0 \cdot 5$–$2 \cdot 0$ cm yr^{-1}. The systematic diverging nature of the motions over the surface of the Earth suggests global expansion, rather than individual peculiar velocities, as the cause. While Carey (1975) in reviewing the expanding Earth hypothesis has referred to evidence indicating a large rate of increase in the Earth's radius, the hot-spot data are clearly in accord with the more modest rate of $0 \cdot 5$–$0 \cdot 6$ mm yr^{-1} inferred from geophysical processes.

For some reason, in discussing geodynamical processes such as those above, many workers assume that compressive forces are needed in the Earth's crust. There is no direct evidence for this, since the number of normal faults and reversed faults mapped are approximately *equal* (Holmes 1965), with no evidence of overall compression or extension in continental blocks. For this reason, there is no need to counter the arguments made by Knopoff (1969) and others that expansion cannot account for compressional lithospheric forces. In any case, the energy used up as *heat* in orogenesis and epierogenesis exceeds that used in pushing parts of the crust by a factor of a hundred (Oxburgh 1967). These examples demonstrate that many of the objections brought against expansion are based on an inadequate understanding of geological and geophysical processes. While none of the evidence is conclusive, there are numerous indications that some expansion has occurred.

6.3. The Rate of Expansion

There are now about 20 determinations of the possible rate of increase of the Earth's radius. When the data (Creer 1967) are plotted as the rate of increase (dR_E/dt) against the age of the data, t (or time span over which the data are averaged), they fall into two main groups. Those older than about 300×10^6 yr form a coherent group in dR_E/dt whose mean is about 0·5 mm yr^{-1}, while values of dR_E/dt for younger data are widely scattered with rates up to ten times this figure. The explanation for this asymmetry is to be found in the fact that the entries in the second group of measurements (with a single exception) are all palaeomagnetic ones, while the first group comprises all other methods (figure 6.1).

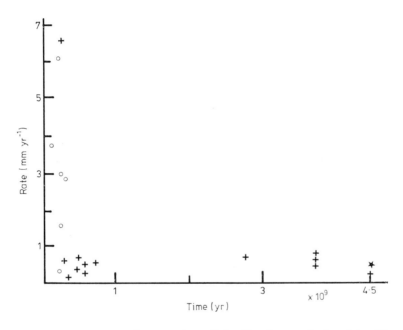

Figure 6.1. Rates of expansion of the Earth as calculated by 20 analyses (figures from Creer 1967) in mm yr^{-1} against the age of the data used to estimate the rate. Circles represent palaeomagnetic methods, and crosses the other methods. The star represents expansion from a completely sial-covered globe $4 \cdot 5 \times 10^9$ years ago.

The palaeomagnetic determination of the Permian radius by Cox and Doell (1961) has the advantage over older methods of not assuming alignment of the rotational and geomagnetic axes, although it does assume a dipole field. The result of Cox and Doell refutes the 45% increase in surface area of the globe at $dR_E/dt \simeq 7$ mm yr^{-1} of Carey (1961; this paper is followed by a reply from Cox and Doell). Egyed (1961) also disagreed with Carey's work, and showed that the method of Cox and Doell is capable of considerable refinement. Van Hilten (1965) used the palaeomagnetic method and obtained a relatively high rate of expansion similar to Carey's. However, since most of the palaeomagnetic work on this subject was done, it has been realised that palaeomagnetism cannot hope to determine expansion rates in the 0·1–1 mm yr^{-1} range, since it is not intrinsically accurate enough. This conclusion was reached by Ward (1963), who used a Fisher analysis and Monte Carlo methods on data of Devonian and younger age. He states that '. . . small inaccuracies in the estimated direction of the Earth's field produce wild fluctuations in the estimated ancient radius . . .' and shows that while Carey's hypothesis of a large radial increase can be ruled out, the palaeomagnetic method is not accurate enough to say anything about Egyed's estimate. The same result is found by Van Andel and Hospers (1969), and further confirmed by Irving (1969), who gives 20% as the threshold of detectability of an increase in radius by using palaeomagnetic data.

The question of the applicability of the palaeomagnetic method was finally and convincingly terminated by Wilson (1970), who analysed 83 high-weight pole positions, and found that the poles fell systematically on the far side of the geographical pole from the source. The most direct interpretation of this is in terms of a magnetic dipole that is displaced 191 (± 38) km northwards along the rotational axis from the Earth's centre. There is also an asymmetry in declination between the northern and southern hemispheres, and Wilson's work shows that ancient dipoles probably moved, and remained asymmetrically placed, for considerable periods of time. The displacement found is not large enough to invalidate general palaeomagnetic results on rocks younger than Tertiary, but is certainly large enough to negate any attempt to test for moderate global expansion.

Thus, disregarding the palaeomagnetic determinations, a group of data remains whose mean is about 0·5 mm yr^{-1}, in approximate agreement with the value of 0·6 mm yr^{-1} expected if the Earth has expanded from a completely sial-covered condition $4·5 \times 10^9$ years ago.

6.4. Causes of Expansion

Regarding the cause of expansion, it seems that speculation has run rife over the years since the idea of expansion first gained recognition in the 1890s. Egyed (1963) has summarised the proffered hypotheses, of which the only ones which can be considered viable are the following: (i) the Jordon–Brans–Dicke theory in which $\dot{G}$ effects cause expansion; (ii) the Hoyle–Narlikar theory in which $\dot{G}$ and/or continuous creation cause expansion; (iii) Dirac's complete theory, in which continuous creation works in conjunction with $\dot{G}$ effects, and (iv) the metric dilation of space–time according to a certain narrow set of properties compatible with metric theories of gravity (Gilbert 1956a, b, 1960, Van Flandern 1975; Chapter 7). A discussion of variable-G expansion is deferred to §6.5.

Continuous creation, as incorporated into steady state cosmologies, is usually conceived as a continuous creation of hydrogen atoms. This hypothesis, if the Earth has expanded from a state with half the present radius, implies that 8/9 of the globe is hydrogen. This is known not to be the case. Taking the rate of continuous creation of matter demanded by the Perfect Cosmological Principle as 10^{-46} g cm^{-3} s^{-1} (Bondi 1952, p143) and the volume of the Earth as 10^{27} cm^3, it is seen that in 10^9 yr, only 3×10^{-3} g of matter would be created inside the Earth if it maintained its present size. Clearly, matter creation proportional to volume fails to account for the expansion.

Matter creation proportional to mass predicts a more favourable rate of 7×10^{-18} g (per g) s^{-1}, but Cohen and King (1969) failed to detect creation at this rate in a laboratory experiment, and set an upper limit of 4×10^{-23} g (per g) s^{-1}. With this rate, and a present mass of 6×10^{27} g for the Earth, about 2×10^5 g of matter would be added to the Earth's mass every second. In $4 \cdot 5 \times 10^9$ yr, the accumulated mass—if this rate of production were maintained—would total only 3×10^{22} g. This is certainly insufficient to account for the bulk of the Earth, even ignoring the fact that the mass created would have been very small early on due to the small mass of the proto-Earth.

What are the prospects for astronomical bodies other than the Earth? Bodies of smaller size, if affected by expansion at a suitably scaled rate, would be expected to show only minor effects (for example, the Moon would be expected to have expanded by something of the order of a kilometre). In the case of the Sun, besides the more manifest effects of expansion, continuous creation could be responsible for the 10^{11} neutrons/ cm^2 s flux postulated by Fowler and Hashemi (1971) to explain the energy source ($n \rightarrow p + \bar{\nu} + e^- + 0 \cdot 78$ MeV) and temperature ($\approx 10^6$ K) of the solar corona. This flux, taken over the solar surface, is of the order of 10^{34} neutrons/s, which is much greater than the 4×10^{11} neutrons/s flux calculated from the creation rate given by Hoyle (1963, p328). Alternatively, using the upper limit set by Cohen and King, and the hypothesis of matter creation proportional to matter, the Sun's mass ($\simeq 2 \times 10^{33}$ g) leads to an expected flux of 5×10^{34} neutrons/s. This is adequate for the Fowler–Hashemi model, which has, however, been criticised by Audouze (1971). The actual flux of protons comprising the solar wind travels at about 500 km s^{-1} with a density of the order of 10 cm^{-3} at the distance of the Earth's orbit (Hundhausen et al 1971). The total flux of protons escaping from the Sun is thus of the order of 10^{35} s^{-1}. Investigations on lunar rocks indicate that the total particle flux has been constant over the last 10^6 yr, at least for that part constituting the medium-energy solar cosmic rays (Lavrukhina and Ustinova 1971). Within the rather large uncertainties attached to it, continuous creation at a steady rate can thus account for the presence of the solar wind, and, should the Fowler–Hashemi model be accepted, also for the dynamics of the corona.

The hypothesis that it is the increase in the number of particles comprising the Earth that causes it to expand may finally be shown to be incorrect when more accurate data on the deceleration of the planet's spin becomes available. An expansion rate of the desired 0·5 mm yr^{-1} could then be accomplished only in conjunction with a decrease in G (with or without an increase in mass), since any tendency for G to decrease causes a modification of the lunar orbit that gives a wider scope to the allowed expansion (Chapter 5). This question awaits refinement of geophysical techniques and estimates of the variation of G, but it should be borne in mind that the separation of $\dot{G}$ effects from global expansion with conventional instrumentation may need a close appraisal of the time scales employed. Atomic time is explicitly used by Hoyle (1972a, b), while Dicke (1957a), Dehner and Hönl (1969), Wilson (1960) and Runcorn (1964) all use the epoch that has elapsed since the origin of the Universe as time coordinate. This is discussed in a different context, in connection with the Dirac and Hoyle–Narlikar cosmologies, in Chapter 2. The various possibilities presented by feasible variations in the particle number, mass and size of the Earth and of changes in G (Wesson 1973) provide a wide region for geophysical speculation.

While it is not a hypothesis resulting in the expansion of the Earth, it is convenient at this point to mention a speculation concerned with the half-life of particles that is in some ways related to continuous creation, and has provoked some relevant geological comments. Kapp has advanced the hypothesis that mass is vanishing from every part of the Earth as a consequence of matter (protons, electrons and nuclei) possessing a finite half-life of rather less than the conventional age of the Earth (Kapp 1960, p245). Atoms are postulated to be created in regions of space devoid of matter, while they 'decay' into non-existence where matter is dense, at a rate governed by this half-life. This means that material bodies such as planets contract at a rate proportional to their densities. The idea of Kapp is hardly acceptable in its basic form, since it contradicts data on particle lifetimes (§4.7), although it might be put on a firmer basis by an appeal to the differing behaviours of inertial and gravitational mass. This possible contradiction of the Weak Principle of Equivalence (Chapter 3) is, however, severely restricted by the results of the Eötvös experiment (Eötvös *et al* 1922), which now has an accuracy of one part in 10^{11} (Shapiro *et al* 1976, Dan'us 1976). Geophysically, the hypothesis has been soundly criticised by Holmes (1965, pp985–6), who observes that radioactive elements and their end products would be annihilated at different rates, making it impossible to reach concordant age estimates by lead, strontium and argon dating, in contradiction with experience. Holmes also points out that if the density of matter in the Earth has been decreasing over geological time, the range of α particles from uranium-bearing rock fragments would have varied. Pleochroic haloes, the distinctive concentric traces left around radioactive inclusions by the emitted α particles, would be observed to be diffuse, which is not in agreement with their sharpness as seen in actual specimens.

The sharpness of pleochroic haloes is an important observation, since it shows that the ranges in matter of fundamental particles have been approximately constant over geological time, which in turn suggests that there have been no changes in their energy of ejection and no fundamental changes in the nature of the parent nuclei. Tests such as this are of great value, and serve to delineate the acceptable and unacceptable geophysical implications of cosmology.

6.5. Variable-G Expansion

The cosmologies mentioned in §6.4 predict various amounts of Earth expansion as consequences of the varying strengths of the dependencies of G on time which they involve. In addition to the release of compaction leading to expansion, there are also other consequences of the $G \propto t^{-1}$ hypotheses that involve expansion in cosmogony.

The effects of a diminishing G on geophysics were pointed out some time ago by Dicke (1957a), who was concerned about the radioactive elements in the Earth and the geothermal flux. At the same time, Dicke discussed the effects of the Brans–Dicke formalism on the radioactive elements, the Sun and the Earth, and stated a belief that the decompression of the Earth caused by decreasing G was responsible for about half of the heat flux reaching the planet's surface ($\simeq 2 \ \mu$cal cm^{-2} s^{-1}). Dicke took the near equality of heat flow over oceans and continents as evidence that a substantial part of the heat flux was not due to radioactive decay in acidic rocks. The heat flow equality is quite possibly the result of a general restructuring of isothermals after the last ice age (Horai 1969), and it cannot be said that there is any real need for a contribution to the heat flow due to varying G. The question is still not settled, since Stacey (1975) believes the heat flow equality to be due to convection under the sea floors with a turnover time of 10^8 yr, with a thermal relaxation time τ_r of approximately the same value in a lithosphere of thickness $(\tau_r \eta_D)^{1/2}$, where η_D is the thermal diffusivity of igneous rocks. Stacey requires that about 3×10^{19} erg s^{-1} (3×10^{12} J s^{-1}) be generated n the core to power the geomagnetic dynamo (precession is often suggested as a driving source, but the heat released by dissipation at the inner/outer core boundary is only $\approx 10^{15}$ erg s^{-1} (10^8 J s^{-1})), and to produce a layer of hot material at the CMB that can escape upwards as pipes. The core heat is supposed to be radiogenic in origin, which implies that there is a notable amount of potassium in an outer core that also contains sulphur, but no account is given of the hypothesis that the decrease in G leads to a lessening of compaction and a release of heat. Stacey's thermal model for the mantle, while plausible in having solidus, adiabatic and actual temperature profiles that keep the lower mantle solid, is rather problematical as regards the CMB layer, which is supposed to give rise to the uprising pipes. The existence of this layer has been criticised by Marsh (1975), along with the concept of hot spots (Lingenfelter and Schubert 1974) as discussed in Chapter 5.

Problems such as these were examined by Murphy and Dicke (1964) following the joint formulation by Brans and Dicke (1961) of the theory now known as the Brans–Dicke scalar–tensor theory (§2.6). In particular, Murphy and Dicke believe that the radioactive heat flux from the Earth's core is too small by itself to drive convection and so allow the generation of a magnetic field. By releasing pressure, due to expansion caused by decreasing G, more heat can be released. There is no basis for this idea in geophysics at the moment, simply because so little is known about the state of the Earth's core, but they did give an estimate of the global expansion expected, which is valid for those theories in which $G \propto t^{-1}$.

The expansion of the planet involved is given numerically by

$$\left| \frac{\delta R_E}{R_E} \right| \simeq 0.1 \left| \frac{\delta G}{G} \right| \tag{6.1}$$

as calculated by Dicke (1962a), and is only 180 km over 4×10^9 yr. This is, of course, completely insufficient to explain the evidence put forward for an expanding Earth in §§6.2 and 6.3, where it seems that an increase in radius by some 3000 km is needed. Murphy and Dicke (1964) used the expansion hypothesis to enable them to adopt a model of the Earth that is chondritic in composition (the apparent heat release of the mantle material is doubled due to decompression caused by decreasing G) and conductive in temperature distribution, a model which agrees with the cold chondrite scheme of Mac-Donald (see Murphy and Dicke 1964, p243). Once again, however, the ignorance prevailing at the moment concerning the constitution of the Earth's mantle must cause one to classify these hypotheses as speculative in content.

The opinion that the expanding Earth hypothesis is plausible on the basis of $G \propto t^{-1}$ is widely held, but Creer (1965b) doubts that a decrease in gravitational force alone is potent enough to cause more than an 800 km increase in radius. He suggests that the 3000 km increase indicated by geological data needs a change over geologic time in ϵ_0, the permittivity of free space. Since $c^2\epsilon_0\mu_0 = 1$ (μ_0 is the permeability of free space), this proposal brings in the awkward possibility of a change in the velocity of light with time. It is quite true that a decrease in G would have no immediately noticeable effects in a laboratory-sized body, because the forces involved in holding it together are not gravitational in nature. However, in a body as large as the Earth, although a decrease in G might not be reflected in a simultaneous expansion, a more or less sudden adjustment would probably take place when the stresses set up reached a critical level. There may therefore be no limit to the expansion, and no need for variations in fundamental parameters other than G. On the other hand, the constant rate of expansion of the Earth (indicated by the data of §6.3) does suggest a dynamic driving force rather than a passive release. Creer suggests that continuous creation of matter, where matter is already most dense, may provide the impetus for the expansion. This hypothesis has affinities with the complete Dirac cosmology.

Jordan (1962a) investigated the geophysical consequences of Dirac's theory, noting (a) the significance of ridges and rifts in the ocean bed; (b) the two well defined levels of the hypsographic curve; (c) the mechanical impossibility of destruction and growth of continents, and (d) continental drift, although (e) not ignoring the fossil evidence for a vanished Gondwana-land, which, however, Jordan would rather interpret in connection with expansion of the Earth from a completely sial-covered state. Jordan takes as evidence for his views the decrease in water-covered areas over geological time, and believes that the Earth was formerly completely cloud-covered, causing the temperature to be 10°C or so all over the globe, with little dependence on location, in the Carboniferous period. This state may have been unstable, the Earth being on the verge of a glaciation that did in fact occur in the following Permian period. The geophysical hypotheses considered by Jordan are not very satisfactory, since he assumes polar wandering, in addition to a constant temperature, to explain the apparent warmth of the Spitzbergen climate in the late Carboniferous, and is obliged to assume that the Pleistocene ice age was of a different nature from that which occurred in the Permian.

The origin of ice ages is probably not connected with the expanding Earth hypothesis or variable G, since glaciation on the analysis of Williams (1975) is a periodic event that occurred not only recently but at epochs of 295×10^6 years ago (Permo-Carboniferous), 445×10^6 years ago (late Ordovician), and 615, 770 and 940×10^6 years ago (Pre-Cambrian). The periodicity of about 150×10^6 yr is difficult to explain on a $\dot{G}$ basis, but rather bears out a hypothesis such as that of McCrea (1975b), who has suggested that the Sun in its orbit around the centre of the Galaxy passes through spiral arms that contain appreciable quantities of dust. The dust, infalling to the Sun, increases its luminosity, so causing the Earth's cloud cover to increase, the higher albedo leading to a drop in surface temperature and the onset of an ice age. The Sun is believed to take 10^6 yr between encounters with the dust clouds in a spiral arm, and 10^7 yr to cross a whole spiral arm. McCrea predicted a mean time between spiral arm passages of about 10^8 yr, which is in agreement with the observed periodicity of the glaciations and a periodicity in the deposition of dust to form lunar soil (Lindsay and Srnka 1975). The hypothesis of McCrea has also been considered by Dennison and Mansfield (1976), who were concerned that the infalling cloud would destroy particles of the order of 1 cm in size within the Solar System, contrary to observation (Hughes 1976). There are several other consequences of the hypothesis, mostly concerned with climatic changes caused by the fact that an encounter with a cloud of fairly modest density would temporarily prevent the solar wind from reaching the Earth (Begelman and Rees 1976). The inference one draws from a consideration of hypotheses such as those of McCrea (1975b) and Weertman (1976, in which glaciations are periodic phenomena caused by variations in solar radiation connected with long-term cyclic terms in the Earth's orbit), is that ice ages should not be used to judge the status of the expanding Earth hypothesis, or of the cosmologies that are compatible with it.

The Hoyle–Narlikar cosmology gives amounts of expansion which are comparable with those of the Brans–Dicke theory, although the adaptability of the former cosmology can allow for an effect of up to about 500 km. Hoyle has reviewed the other geophysical and cosmogonical effects predicted from the cosmology of Hoyle and Narlikar (Hoyle 1972a, b, Darius 1972), but an attempt to use data on the lengthening of the day and orbital motion of the Moon does not result in any useful outcome, since the data are too inaccurate. Geophysically, some of the statements made by Hoyle and Narlikar (1972b) are erroneous. While they are correct in pointing out that the Earth's crust must be continuously at the limit of its strength as G decreases, due to the ever-present tendency to expand, this has little of the implied relevance to the Palaeozoic continental drift question. The continents are indeed slowly pushed apart, but the Hoyle–Narlikar theory (a) does not give sufficient expansion to explain the origin of oceans, as implied, since this can only be accomplished by a 3000 km expansion. With regard to continental drift phenomena such as the postulated northward drift of India into Asia, or the break-off of Australia from Antarctica, the Hoyle–Narlikar theory does not explain (b) why the stresses caused by expansion should be directed horizontally; (c) why the stresses carry on pushing continents over vast distances when their influence should be purely local; (d) why Creer (1965a, b) was able to fit all the continents onto a globe half its present size; (e) why the sea-floor spreading rate (§6.2) is an order of magnitude larger than the rate given by the expansion of the Hoyle–Narlikar theory, and (f) why it should still be conceived that sial (continents) can plough through sima (oceans) with little deformation, when sima is stronger than sial. The last objection is one that has now been realised, after decades of reiteration, and circumvented by the hypothesis of plate tectonics, but the other objections to the Hoyle–Narlikar theory, as applied to geophysics, still remain.

The result of the Hoyle–Narlikar and Brans–Dicke cosmologies that G decreases with time was quickly taken as a basis for the expanding Earth idea and related hypotheses. The most notable proponent of this view has been Egyed, who has also proposed that the origin of the Solar System can be explained on the basis of $G \propto t^{-1}$ (Egyed 1960). This latter is an elegant hypothesis in which planets are imagined to be thrown off, originally in the form of annular rings, as the Sun expands, the radius at any time being taken as $R_\odot \equiv (R_0 + \beta_\odot t)$, which increases linearly with time (R_0 and $\beta_\odot$ are constants). By equating the surface gravity of the Sun to the centrifugal force $R_\odot \Omega_\odot^2$ for an element of the surface, it is seen that matter will escape from the Sun when

$$R_\odot \Omega_\odot^2 = \frac{GM_\odot}{(R_0 + \beta_\odot t)^2} = \frac{\text{constant}}{t} \frac{M_\odot}{(R_0 + \beta_\odot t)^2}. \qquad (6.2)$$

Planets might be expected to form from rings shed repeatedly by the Sun as it expands in this fashion, always assuming that G is decreasing fast enough to enable matter to become decoupled from the Sun against the natural tendency towards an overall decrease in rotational velocity and a consequent reorganisation of binding energy. Further investigation shows that the Titius–Bode law can be accounted for, but the theory has been criticised on the grounds that the Sun should be left in a state of rapid rotation, contrary to observation (Williams and Cremin 1968). In any case, $G \propto t^{-1}$ theories do not seem to be able to explain the apparently observed expansion in geophysics (since none of them give 3000 km of expansion over $4 \cdot 5 \times 10^9$ yr in the Earth), so the relevance of these theories to cosmogony also seems doubtful.

Continuous creation, conceivably coupled with a time-variable G as in the complete Dirac theory, might account for the required expansion if the need for this is accepted on the data noted above. The continuous creation hypothesis is interesting, but within modern theories of big bang cosmology there is no need for it. The related proposal that it is the space which expands in general relativistic cosmologies has not been in favour for some time. It harks back to Sir James Jeans'

7

much-quoted comments, the best known of which is that the form of spiral galaxies suggests that matter is here being poured into our Universe from another, extraneous dimension (Jeans 1928). It can, however, be put on a firmer basis within general relativity by using the work of Gilbert (1956b), or as indicated in Chapter 7. MacDougall *et al* (1963) have compared universal and terrestrial expansion, and have shown that when the Earth's radius is substituted for R in the Hubble recession formula (velocity $= RH$) where H is taken to be 75 km s^{-1} Mpc^{-1}, as indicated by astrophysical data, the resulting predicted expansion is 0·5 mm yr^{-1}, which is a very acceptable result.

6.6. Ancient Gravitational Acceleration

Besides evidence for an expanding Earth, there has been much activity in the field of palaeogravity with the object of studying long-term trends in g_E—the local acceleration due to gravity at the surface of the Earth—by means more restricted than whole-planet geophysics. Data on g_E can be used to obtain limits on $\dot{G}$ and ancient values of R_E. Experiments that have been proposed for g_E include investigations of (i) the compaction of sediments by overlying loads; (ii) the growth of ancient salt diapirs (not now regarded as feasible since the growth rate depends mainly on local sedimentation conditions); (iii) the muscular development of Palaeozoic animals; (iv) the depth of grooves in basement rock caused by boulders trapped in moving ice sheets or ancient glaciers, and (v) various artefacts related to the last suggestion but involving animal footprints in mud beds, boulder impressions, and similar effects.

Stewart (1972) looked at the compaction of sediments (method (i) above) with the intention of testing the expanding Earth hypothesis by measuring the ancient value of g_E (which should have decreased on a simple expansion model). To detect the expected one part in 10^9 yr^{-1} change needs data covering a long span of time, since g_E is varying temporally on a time scale of decades at present. Faytel'son reports changes of one part in 10^7 yr^{-1} in recent decades over the USSR due to geological processes (see Stewart 1972, p322). To average out such effects, Stewart measured the consolidation of the London clay as produced by the overlying layers. Unloading by erosion of this stratum began 26×10^6 years ago, but data on the lithification of Plio-Pleistocene clays and oceanic ooze show that compaction measures are liable to be extremely variable. The result of measurements on the London clay only show the rather rough result that g_E (26×10^6 years ago) was within a factor two of g_E(present). It is thus probable that g_E has not altered in recent geological time, and the 'exhumation' method (that is, comparing a sedimentary layer that was once deeply buried with its eroded form) indicates the limit of variation as being $\lesssim 6$ parts in 10^8 yr^{-1} over the last 26×10^6 yr. The limit set by the compaction method gives a somewhat erroneous impression of exactitude. One can never eliminate all other geological influences from such investigations, and the results are therefore not capable of a completely definitive interpretation.

In view of the inconclusiveness of many of the tests aimed at measuring ancient values of g_E and G, it is necessary to find a method that is, in the first place, of a nature in which statistical methods can be used to achieve useful accuracy and, in the second place, is also capable of measuring changes in the radius of the Earth (Wesson 1973). The local gravitational acceleration, g_E, on the surface of the Earth measures both G and R_E, since $g_E = GM_E/R_E^2$. Measuring ancient values of g_E is, by conventional standards, inexact and would give no very close limits on G, but variations in R_E obviously produce a much more notable effect on g_E. Potentially, perhaps the most accurate way to estimate g_E in the past lies in examining the ejecta from ancient volcanoes, since, although the angle of repose of ejected material is not dependent on g_E, its lateral distribution is. The range of particles ejected from a volcano with speed v_e, at any angle β_e from the horizontal, is given by the elementary relation

$$\text{Range} = \frac{v_e^2 \sin 2\beta_e}{g_E}. \tag{6.3}$$

While quite detailed geological maps of ancient volcanoes are available, this test now appears to be one which would be difficult to apply, as has been pointed out by N W Molyneux (personal communication). (i) Meteorological conditions would have been different at different times during the eruption of a Palaeozoic volcano and also different at different altitudes; (ii) the estimation of the ejection velocity would be problematical, as would (iii) the measure of the horizontal range to sufficient accuracy, and (iv) corrections for purely local variations in the gravitational acceleration. These objections to the proposed test, while serious, could be circumvented if a statistical method were employed to find the parameters necessary to estimate G and/or R_E from equation (6.3), using a large number of ejected objects chosen to have the same size and shape. The attainable accuracy could be as good as other proposed ways of evaluating ancient G, and might be considerably better as a test of changes in R_E.

Rosenberg and Runcorn (1975), in summarising the import of a large number of biological studies of the history of the Earth's rotation as monitored by growth rhythms in corals, point out that the proposal of Wesson (1973, 1975b), that expansion may be occurring without changes in G, could be tested. Ancient values of g_E could be calculated using the technique of Creber (1975), who has reported on ways in which the growth of ancient wood can help to determine past gravity. Some way of determining ancient values of g_E, G or R_E without going through an analysis that involves the rotation of the Earth (cf the methods discussed in Chapter 5) would be of enormous help.

It should be noted here that slightly misleading statements are sometimes made with regard to the subjects of expansion and rotation. Notably, Müller and Stephenson (1975) have re-analysed ancient eclipse observations to estimate the deceleration of the Earth's spin, and have found that modern, mediaeval and ancient data of all kinds indicate approximately constant lunar and Earth decelerations, contrary to conclusions reached by other workers such as Newton (§§5.3 and 5.4). What they call the non-tidal acceleration has also been constant and positive over the historical period, and could be due to sea level changes and/or geomagnetic effects. The *overall* LOD change is, of course, negative, at $-2 \cdot 5\,(\pm 0 \cdot 3)$ ms century^{-1}, in agreement with coral data. Confusion can arise, however, because of loose nomenclature. Disregarding the non-tidal term is assumed implicitly by Müller and Stephenson to leave

188

a tidal term which can be identified as tidal because it affects both the Moon and Earth. This is not necessarily so, because expansion of the Earth and the lunar orbit (as suggested in this and Chapter 7) would also affect both parameters, and would mean that part of the 'tidal' term is in fact due to expansion. The results of Müller and Stephenson would then show that the tidal deceleration and the expansion rate have been constant.

Apart from the direct effects of expansion of the Earth and its influence on g_E and rotation, there are certain aspects of geophysics (Wesson 1973) that have the uncomfortable appearance of being coincidences, but which might find explanations in cosmological processes. One such process, the continuous creation of matter (§6.4) at a rate depending on the gravitational potential, was proposed as a driving force for expansion with a view to maximising the creation rate at the centre of the Earth. It would seem to be difficult to affect the potential at the centre by more than a factor of three compared with the surface by adjusting geophysical parameters, as pointed out by Straker. It was noted first by Benfield, and commented on by Jeffreys (1970, p202), that the acceleration of gravity g *inside* the Earth is roughly a constant throughout the thickness of the mantle. Straker has looked more closely at this phenomenon. If the density of the mantle is $\rho_m(r)$, then $g(r)$ is given by

$$g(r) = \frac{G}{r^2} \int_0^r \rho_m(r) \, 4\pi r^2 \, dr, \qquad (6.4)$$

which can be implicitly differentiated and the derivative put equal to zero to obtain the condition for $g(r)$ to be constant. This results in

$$\rho_m(r) = \frac{g(r)}{2\pi G r} \sim \frac{1}{r}. \qquad (6.5)$$

This relation is good to about 10% accuracy in the mantle (Jeffreys 1970), and so the constancy of $g(r)$ would seem to be a consequence of the Earth's composition rather than of the operation of any cosmological process. It has some interesting implications with regard to expansion of the globe, as can be seen from considering the corresponding pressure–depth equation. This is given in terms of the pressure at the surface

7*

189

(R_E) by

$$p(r) = \int_r^{R_E} g\rho \, dr = g\rho_0 \ln \left(\frac{R_E}{r} \right) \qquad (6.6)$$

where constancy of g has been assumed and ρ_0 is a constant. Changes on a secular time scale in R_E and possibly in G are seen to bring in the question of the secular stability of the pressure–density relation in the Earth. The question of the geochemical evolution of a planet under conditions of a variable G with a changing radius and density is one that could profitably be pursued in detail.

6.7. Laser Ranging and Satellite Tracking

A device that can directly detect motions of expansion is available in the form of the lunar laser reflector (Alley *et al* 1969). This instrument, in theory, can determine the Earth–Moon distance to an accuracy of 15 cm, and with the accumulation of many years' data it is feasible to determine directly the rate of evolution of the lunar orbit and, by surveying techniques, of motions in the Earth. A summary of the lunar laser experiment has been given by Bender *et al* (1973), and improved instrumentation may well enable an accuracy of 3 cm to be achieved over a period of several years' observation. On data over the period 1969–1975, an accuracy of about 40 cm was reached in terms of the radial error bracket for tracking of the Moon's orbit. Two groups have analysed these data, as summarised by Dan'us (1976), and have put limits on the influence of the Nordtvedt effect in the Earth–Moon system; that is, the Earth's gravitational self-energy was found to contribute equally to its inertial and passive gravitational mass to an accuracy of $\pm 3\%$ (see Chapter 3). These analyses had the object of detecting an oscillation in the Earth–Moon distance with a period of one month. They confirm the results of Chapter 2 in putting a limit on the coupling parameter ω_{BD} of the Brans–Dicke theory of $\omega_{BD} \gtrsim 29$, implying that no notable departure was found for the effects expected from general relativity alone. Since other influences (such as tides) that lead to a uniform recession of the Moon from the Earth were removed from the analyses, it is still plausible that the

190

Moon's orbit may be expanding due to some cosmological process that only affects it in a uniform manner. It is also plausible that there could be a term present due to a uniform expansion of the Earth, and improved instrumentation may allow future detection of any such geophysical movement.

The problem would then be to separate the effects of Earth expansion from those of continental drift, plate tectonics and polar wandering. The International Latitude Service has been monitoring apparent polar motions for some time, but these data are not accurate enough to detect lateral movements on the Earth's surface at the expected circumferential extension rates predicted from the expanding Earth hypothesis, or expected from plate tectonics. Nevertheless, some authors (see, for example, Proverbio et al 1972) are inclined to believe that they can detect drift rates of 1–2 cm yr^{-1}. This they have concluded by using data on changes in latitude, and have deduced movements of continental blocks, by observations of time scale discrepancies, in the sense of a westward movement of the Earth's crust corresponding to a deceleration in the spin (Proverbio and Poma 1975). This interpretation would seem to be doubtful because of the low level of correlation being sought for in problematical data based on time scale fluctuations (O'Hora 1975, Morrison 1975). Changes of position of the type expected from plate tectonic models are inferred from such time scale fluctuations, and these are affected by changes in the Earth's spin rate and the instantaneous orientation of the pole. The effects of irregular winds, in particular, cause fluctuations on a time scale of months and involve torques of the order of 10^{26} dyn cm ($= 10^{19}$ N m). For comparison, the seasonal wind changes with a 1 yr time scale involve torques $\simeq 3 \times 10^{25}$ dyn cm, while the fluctuations thought to originate at the CMB involve torques of $\simeq 4 \times 10^{24}$ dyn cm and mostly have much longer time scales of 30–300 yr. It is therefore difficult to measure continental movements (or related effects) using only time scale data. The separation of the secular motion of the pole from motions such as those due to continental drift has been discussed by Mueller and Schwarz (1972) with special attention to very long baseline interferometry, but this technique has yet to be extensively employed in geophysics.

An instrument of similar interest is the time-varying gravitational satellite proposed by Groten and Thyssen-Bornemisza (1972), since a satellite in an equatorial, almost circular orbit would be best suited to estimating $\dot{G}/G$. Changes in the Earth's volume, as expected on the expanding Earth hypothesis, would not sensibly affect the semi-major axis of the satellite (and neither would a secular change in sea level). Hence, the effects of a change in G and a change in I (the mean moment of inertia) could be separated, providing a check on the results of Weinstein and Keeney (1973a; §5.3). The accuracy of Doppler determinations of station positions using satellites (Anderle 1972) does not give as high an exactitude to the determinations of interest as does the lunar laser reflector, so it would seem that continuous laser tracking represents the best possibility of directly testing for cosmological effects within the Earth–Moon system.

If laser ranging and refined geophysical data in favour of expansion (§6.2) give an expansion rate in agreement with the geophysically implied rate of 0.5 mm yr^{-1} (§6.3), the causes of the expansion (§6.4) will be narrowed down considerably, since the simple $G \propto t^{-1}$ cosmologies will not suffice to account for it. The universal Hubble law of MacDougall *et al* (1963) will then be strongly supported, as will the Dirac theory. The former involves no dependence of G on time, while the latter does, and the most direct way to differentiate between the two may be through estimates of the ancient local gravitational acceleration (§6.6). Emphasis must therefore be very clearly directed towards improving the observational and theoretical grounds for delineating the differences between the variable-G hypothesis and the expansion hypothesis.

7. *Expansion in Relativity*

7.1. Introduction

Attempts to detect a possible time-dependence of Newton's gravitational parameter G were unsuccessful until the apparent discovery by Van Flandern (1975) that $\dot{G}$ is significantly finite, with $\dot{G}/G \simeq -8 \times 10^{-11}$ yr^{-1} (see §5.8). This has been discussed by him on the basis of three hypotheses: **G** (that G depends on atomic time), **M** (that cosmic bodies are all losing mass on astronomical time scales), and **S** (that the space–time continuum is expanding locally). In view of the implausibility of **M** and the lack of theoretical justification for **S**, Van Flandern interpreted his result in terms of **G**. This supports certain cosmological models such as the Hoyle–Narlikar mass field concept and, to a lesser extent, the Brans–Dicke scalar–tensor theory (Chapter 2). The purpose of the following is to examine to what extent hypothesis **S** might be taken as an alternative explanation of the facts.

There is justification for this from several directions. If one could show that Van Flandern's results could be explained without involving a changing G, such an explanation would be preferable to a variable-G one even at the expense of adopting a hypothesis as remarkable as **S**. Further, there is direct support for the latter type of idea from the data of Chapter 6. While there may be no theory available that yields **S** as a direct feature of its formulation, it is quite plausible (§2.8 and Chapter 3) that a metric theory of relativistic gravitation can be derived that is compatible with it. Indeed, the de Sitter model of Einstein's general relativity is an example of a (somewhat trivial) model that evolves in accordance with **S**. It is therefore within the bounds of legitimate speculation to ask what conditions should a metric theory of gravity possess to ensure that it yields dynamical results in accordance with **S**?

193

Section 7.2 is in the nature of a speculation, forwarded with the object of answering the above question. In fact, one is far from being totally without clues in this matter. Many self-consistent theories of gravity are gauge theories, and it is for this reason that these theories (for example, the Brans–Dicke, Hoyle–Narlikar and complete Dirac theories) have some frame in which the equations of the theory are formally the same as those of Einstein (cf §2.8 on gauge theories). One may therefore assume that there is some frame or gauge in which the equations are those of general relativity. Particular solutions derived using these equations, as for example cosmological models and their metrics, will also be the same as for Einstein's theory in that gauge. (Hoyle and Narlikar (1974) use this correspondence to take over the formal form of the Friedmann models from Einstein's theory to the mass field theory.) Another clue to finding the conditions under which S might hold is given by the restrictions that hold on the metric tensor g_{ij} in Einstein's theory. The tensor g_{ij} and some of its derivatives are continuous in general relativity, and such properties embody certain requirements about the nature of space–time that are of interest as far as S is concerned. To be specific, in general relativity it is possible to find a coordinate frame in which g_{ij} and $g_{ij,\,k}$ and $g_{ij,\,k\alpha}$ are continuous in the space–time, even across matter–vacuum interfaces (Roman indexes $= 1$–4 and Greek $= 1$–3; Synge 1966, Robson 1972; §3.4). The metric continuity property and the possibility of choosing a gauge in which the equations are formally the same as those of Einstein will be seen to provide an adequate basis for deriving the conditions under which Van Flandern's hypothesis S can be valid. The same conditions will show that the effects of the alternative hypothesis M are negligible, and G will be able to be taken as a true constant.

It should be emphasised that despite the above comments one is not dealing with general relativity in §7.2. It will be found that the conditions required for S, while physically reasonable, are not in general those holding in Einstein' theory. The contents of §7.2 are a speculation (albeit quite a concrete one) on a metric theory of gravity whose equation are those of Einstein in some special gauge.

In the other sections of this chapter further consideration i

194

given to matters suggested by §7.2, but in §7.3 the discussion is again confined to models that are derivable within the framework of Einstein's theory, and in succeeding sections an attempt is made to limit the discussion to those avenues that might merit further investigation.

7.2. $\dot{G}$ and Expansion

As noted above, for the purposes of this section only, one wishes to assume the existence of a metric theory which possesses a gauge that allows one to use Einstein's equations. With this assumption (A), one hopes to be able to obtain simple relations that are valid in the relativistic regime and can be used to examine the status of Van Flandern's S hypothesis. Solutions of Einstein's equations (3.1) where Λ is finite,

$$R_{ij} - \frac{R}{2} g_{ij} + \Lambda g_{ij} = \frac{8\pi G}{c^4} T_{ij}, \qquad (7.1)$$

nearly always result in relationships too complicated to be useful. If, however, one employs the spherically symmetric metric

$$ds^2 = c^2 e^{2\phi} \, dt^2 - e^{\lambda} \, dr^2 - R^2 \, (d\theta^2 + \sin^2 \theta \, d\phi^2) \qquad (7.2)$$

equation (7.1), where $\Lambda = 0$, can be made to reduce to five certain relations:

$$\frac{2GM}{c^2 R} = 1 + \frac{e^{-2\phi} \dot{R}^2}{c^2} - e^{-\lambda} \left(\frac{\partial R}{\partial r} \right)^2 \qquad (7.3.1)$$

$$\dot{M} = -\frac{4\pi R^2 p \dot{R}}{c^2} \qquad (7.3.2)$$

$$\frac{\partial M}{\partial r} = \frac{4\pi R^2 \epsilon}{c^2} \left(\frac{\partial R}{\partial r} \right) \qquad (7.3.3)$$

$$\frac{\partial \phi}{\partial r} = -\frac{\partial p/\partial r}{p + \epsilon} \qquad (7.3.4)$$

$$\dot{\lambda} = -\frac{2\dot{\epsilon}}{p + \epsilon} - \frac{4\dot{R}}{R}. \qquad (7.3.5)$$

Here, M is the total mass interior to radial comoving coordinate r where matter has pressure p and energy density ϵ (Podurets

1964, Misner and Sharp 1964). The coefficients ϕ and λ are functions of r and t, and R is not constrained to be comoving: $R = R(r, t)$. Conventional units ($G \neq 1$ and $c \neq 1$) are kept for ease of numerical computation.

In addition to assumption (A) above, one also needs an assumption that relates more directly to hypothesis **S** and the metric tensor. One is fairly certain that **S** can be accommodated into a metric theory by a condition on the g_{ij}, since what **S** is saying is that all bodies are in some way affected by the expansion of the background cosmological metric. What is needed is a way to couple the expansion of the cosmological background to a condensation present in this background. Conditions on g_{ij} in general relativity are often used in situations of this sort (Ne'eman and Tauber 1967, Cahill and Taub 1971), but these conditions are not strong enough for present needs because they allow one to carry out coordinate transformations on one or both sides of a boundary that can be used to imbed a static sphere in a moving background. This aspect of requiring continuity of the g_{ij} and its derivatives in only some frames and not all is not compatible with **S**. One is thus obliged to go outside Einstein's theory and assume (B) that there exists a condition on the metric and its derivatives that makes $\dot{R}/R = \dot{S}/S$ for a body of radius R embedded in a cosmology characterised by scale factor S (where $R \equiv Sr$). The latter assumption (B) is really a mathematical statement of Van Flandern's formulation of **S**, and an examination of equations (7.3) shows that it can be achieved by a continuity condition on g_{ij} and its derivatives as noted.

The assumptions (A) and (B) are the main ones required to obtain **S**. To be exact, the result $\dot{R}/R = \dot{S}/S = H$ (where H is Hubble's parameter for the cosmology) also requires some ancillary assumptions which one can collect as follows: (C) the pressure p must be small compared with the energy density ϵ (a condition satisfied in all except collapsed bodies, where $p = \epsilon/3 = \rho c^2/3$ is the limiting configuration of a Fermi sphere); $\dot{R}$ must be assumed to be not identically zero; the Universe exterior to the condensation must have a reasonably behaved, locally flat metric of the type provided, for example, by the Robertson–Walker models; and the matter in the cosmological field has to be comoving. These last ancillary conditions (C) are not as important as (A) and (B) above, but as may be checked by using equations (7.3), they ensure that the desired result $\dot{R}/R = \dot{S}/S$ follows once (A) and (B) have been accepted.

Given the three assumptions (A), (B) and (C), the surface of a local object expands with the Universe at a rate governed by the cosmology. To recap, this depends on being able (A) to use Einstein's equations in some gauge but with (B) a non-Einsteinian metric condition applied. The other set of assumptions (C) are physically reasonable and are expected to be obeyed in any well founded theory. A more detailed investigation would be needed to see just how sensitive the conclusion $\dot{R}/R = \dot{S}/S$ is to the noted assumptions, but from a physical point of view the system considered above is not in any sense exceptional. All the assumptions entail is the statement that motions on a time scale of the evolution of the metric (that is, H^{-1}) are different in nature from the faster motions one is familiar with in local astrophysics (a similar result was obtained by Gilbert; see §2.5). This does not mean that systems cannot collapse or contract. All processes with time scales much shorter than H^{-1}, such as the Newtonian gas dynamics of collapsing clouds, proceed as usual, while the hypothesis says nothing about the formation of galaxies, etc, in the early Universe when p was not negligible. In summary, given the conditions (A)–(C) above, a matter sphere expands with the Universe and its velocity of expansion obeys Hubble's law with the same Hubble parameter $H = \dot{S}/S$.

This statement of hypothesis S can be tested empirically and immediately. Firstly, consider Van Flandern's result of $\dot{G}/G = -8 \times 10^{-11}$ yr^{-1}, which is based on lunar occultations and uses Atomic Time. Contributions to an apparent $\dot{G}$ caused by tidal slowing of the Earth's rotation and tidal secular acceleration of the Moon's motion do not affect the final result (contrary to previous attempts to detect $\dot{G}$ by misplaced usage of Ephemeris Time, which is determined by the motions of the planets). Vinti has shown in equation (5.1) that if a planetary semi-major axis (l) changes at a rate $\dot{l}$ in the classical two-body problem, then the dynamical effects are equivalent to a time-dependent G with $\dot{G}/G = -\dot{l}/l$, and vice versa (Vinti 1974). For the Earth–Moon system, hypothesis S predicts $H \equiv \dot{R}/R = \dot{S}/S$ (cosmological) $\simeq 75$ km s^{-1} Mpc$^{-1} = 2\cdot5 \times 10^{-18}$ s^{-1}. The $\dot{G}$ value, interpreting Van Flandern's result in terms of **S**, is $H=\dot{l}/l=|\dot{G}/G|=2\cdot6\times10^{-18}\text{ s}^{-1}$.

Secondly, consider the Sun and relation (7.3.2) on the basis of **S** with $\dot{R}/R=2\cdot5\times10^{-18}\text{ s}^{-1}$. $R=R_\odot=7\times10^{10}$ cm is known, as is the mean pressure $\langle p\rangle=\langle p_\odot\rangle\simeq10^{17}$ dyn cm^{-2} (10^{16} N m^{-2}; Cox and Giuli 1968). Such a value for $\langle p\rangle$ does not contradict the condition $\langle p\rangle/c^2\ll\langle\rho\rangle$. The predicted mass loss rate is then $|\dot{M}_\odot|\approx10^{12}$ g s^{-1}. This is the same order of magnitude (Vinti 1974; §6.4) as the observed 10^{11}–10^{13} g s^{-1} that is being lost from the Sun in the solar wind.

Thirdly, there is the $\dot{R}_{\rm E}=HR_{\rm E}\simeq0\cdot5$ mm yr^{-1} coincidence of §§6.5 and 7.4, which is now explained.

The above three tests presuppose that the use of hypothesis **S** remains valid down to the sizes of bodies comprising the Solar System, and applies to their orbits. The validity of this approximation is difficult to judge, but if Van Flandern's **S** hypothesis is valid, it should also hold in more readily testable astrophysical situations like clusters of galaxies, which could thus be expanding with effective Hubble parameters $H\simeq75$ km s^{-1} Mpc^{-1}. Observational data are not yet exact enough to refute the expansion of clusters at $\simeq100$–200 km s^{-1}, but work is in progress to confirm or deny them. In geophysics, there is some data in agreement with **S** that has long been regarded as curiously coincidental (Wesson 1973). In retrospect, the consistency of **S** can be checked via equation (7.3.2) in showing that hypothesis **M** is of no significance, since $\dot{M}/M\approx(\langle p\rangle/\langle\rho\rangle c^2)(\dot{R}/R)$ so that the mass loss time scale for any body of mass M is $T_{\rm M}\approx(c^2\langle\rho\rangle/\langle p\rangle)H^{-1}$, where $\langle\rho\rangle$ is the mean matter density. All bodies except degenerate Fermi spheres or collapsed stars have $\langle p\rangle/\langle\rho\rangle c^2\ll1$, so the mass loss time scale is enormously longer than the expansion time scale H^{-1}.

It is therefore seen that the hypothesis of local space–time expansion (**S**) is consistent with a metric theory of gravity which possesses (A) a gauge in which Einstein's equations are valid and in which the assumptions (B) and (C) noted above can be taken to hold. Under these assumptions, hypothesis **M** is shown to be unimportant and G is a true constant. The expansion of all systems is not a new idea but has been considered in physics for a long time, notably by Gal-Or (1971) in connection with the bases of cosmological thermodynamics.

Dynamically, hypothesis **S** *can* be interpreted in terms of hypothesis **G** with $\dot{G}/G = -H$, where H is Hubble's parameter. This explains Van Flandern's result $\dot{G}/G = -8 \times 10^{-11}$ yr^{-1}. All effects of hypothesis **S** are of the same order of magnitude as those of variable-G theories in which $G \propto t^{-1}$, and so are very hard to detect. Astrophysical systems provide many examples of cases in which metric continuity could be used to calculate the consequences of hypothesis **S** with the best hope of observational validation. In these cases, hypothesis **S** does not contradict Newtonian dynamics, but says that such processes are basically different from those on a scale in which the whole geometry of the Universe is changing. This feature sets the hypothesis considered in this section, notably via assumption (B), apart from Einstein's theory, but the departure is not more drastic than that involved in (say) the Brans–Dicke theory, and it is not unreasonable to suggest that some process may be operating in nature that constrains systems to set up physical conditions that allow **S** to operate as outlined roughly above.

7.3. Expanding Scale-Free Cosmology

While the speculation of the previous section is suggestive of how one might explain Van Flandern's result without a changing value of G, one would like to gain some indication of the consequences of such ideas as suggested above in order to judge the acceptability of the component assumptions. While (C) is unremarkable, the assumption (B) might be expected to have serious consequences for the large-scale properties of a space–time obeying it, and the matter content of such a space–time should not be subject to unreasonable constraints if the assumptions are to be considered as feasible ones. In this section, therefore, the large-scale astrophysical consequences of a space–time that might be compatible with **S** will be briefly derived in the form of a cosmology.

While (B) suggests an interesting property for a space–time, in this section the discussion is confined to a weaker form of it that is contained within conventional theory and which can therefore be judged in a more definite way than the account of §7.2. This means that one can return to Einstein's general relativity, the object being to obtain a model for a cosmology that possesses the property of not admitting coordinate transformations to be carried out in it. General relativity is, of course, unaffected by the coordinate frame anyway, but the desired restriction is not a trivial one. It means that *the model should not contain scales of length or time* that allow the coordinates r and t to be made dimensionless in equations (7.3) and so allow dimensionally consistent coordinate transformations to be carried out. In other words, one does not want any quantity with the scale of a length in the cosmology (this automatically rules out quantities with the dimension of time, formed by multiplying a possible length scale by the inverse of the velocity of light, c). The absence of scales would in any case appear to be a natural postulate to use as the starting point for deriving a model that is to explain the structure and dynamical behaviour of the Universe, and a statement of the absence of scales is one means of implementing the Cosmological Principle into such a model.

There is a simple way to ensure this property, known as self-similarity. The method has been but little used in relativity, although it is of established significance in Newtonian physics and is known to lead often to inhomogeneous distributions of the matter in systems that obey it. Flows of matter far from boundaries tend to settle down into self-similar flow regimes, and this is another way of saying that a cosmology ought to be described by a model of the self-similar, scale-free sort. Several collapsing local solutions of this type have been studied (Cahill and Taub 1971; see also Barenblatt and Zeldovitch 1972, Disney *et al* 1969, Vitello and Salvati 1976, Barnes and Whitrow 1970, Thompson and Whitrow 1967, 1968, Datt 1938, Oppenheimer and Snyder 1938, McVittie 1956, Penston 1969, Hunter 1969, McNally 1964, Sedov 1959), but the idea has not been widely applied in cosmology.

The results to be quoted below are derived in Henriksen and Wesson (1978), who discuss general relativistic self-similarity in detail. It is relatively easy to confirm the following conclusions by resubstitution into equations (7.3), and reference may be made to Barenblatt and Zeldovitch (1972) for a review of the self-similar technique.

The self-similar approach to equations (7.3) consists in writing the dimensional unknowns p, ϵ, M and R as products of two functions, one of which is dimensional and the other dimensionless. The first has to be written solely in terms of one of the variables r or t, and the constants entering the problem (that is, G and c; one variable is enough because r and ct, for example, are dimensionally equivalent). The second function expresses how the quantities, p, ϵ, M and R vary in space and time, but because it must be dimensionless, this function depends only on ct/r (or a power of it, but this adds nothing new). When one solves equations (7.3) subject to this constrained way of expressing the unknowns p, ϵ, M and R, one obtains ordinary as opposed to partial differential equations in these four quantities, plus the metric coefficients ϕ and λ. The latter, which are dimensionless in a metric of the form of equation (7.2), are only functions of ct/r. The reduction to ordinary differential equations happens because the arbitrary functions of r and t that one would normally encounter in solving equations (7.3) have become simply constants (since all the dependence on ct/r has been absorbed by the way of writing the unknowns discussed above). The most straight-forward way to implement the procedure outlined above is to make the definitions

$$\epsilon \equiv c^4 \epsilon_* / 8\pi G r^2, \qquad p \equiv c^4 p_* / 8\pi G r^2, \qquad M \equiv c^2 r M_* / 2G$$

and $$\hspace{8cm} (7.4)$$

$$R \equiv r R_*.$$

The numerical factors here are introduced purely for conveni-ence. The starred quantities have to be functions only of the dimensionless variable ct/R. When one substitutes these specifications into equations (7.3), one can solve the resulting equations for ϵ_*, p_*, M_*, R_*, ϕ and λ, and so find ϵ, p, M and R in explicit form. Perhaps the simplest, evolving scale-free class of solutions to Einstein's equations that one can derive in this way are those that have zero pressure ($p_* = 0 = p$; Henriksen and Wesson 1978). The complicated equations (7.3) then reduce to a very simple set with $e^{2\phi} = 1$, $e^\lambda = R_*^{-4} \epsilon_*^{-2}$, $M_* = $ constant and $\epsilon_* = M_* R_*^{-2} (R_* - ct R_*'/r)^{-1}$ where the prime denotes a derivative with respect to the dimensionless variable ct/r. The parameter R_* is given by solving the equation $R_* R_*'^2 + (1 - M_*^2) R_* - M_* = 0$ which is obtained from equation (7.3.1). When $M_* = 1$, a special solution is obtained which is very simple, and is given in explicit form by

$$
\left.
\begin{aligned}
p = 0; \qquad & \epsilon = \left(\frac{c^4}{18\pi G a^2 r^2}\right) \frac{1}{(1 + ct/ar)(1 + ct/3ar)} \\[2mm]
M = \frac{c^2 r}{2G}; \qquad & R = r\left[\frac{3}{2}(a + ct/r)\right]^{2/3} \\[2mm]
e^{2\phi} = 1; \qquad & e^{\lambda} = \frac{(3/2)^{4/3}(a + ct/3r)^2}{(a + ct/r)^{2/3}}
\end{aligned}
\right\}
\qquad (7.5)
$$

where a is an arbitrary constant. It is seen that the model starts off at $t = 0$ with an inverse-square density profile $\rho \propto r^{-2}$ and evolves to a homogeneous distribution as $t \to \infty$ ($\rho \propto t^{-2}$). It always retains, however, the inhomogeneous property of having $m = m(r) \propto r$, meaning that the density is in general inhomogeneous. The model in equation (7.5) is the parabolic $(2GM/c^2 r = 1)$ case of a class of solutions, some of which expand forever and some of which recollapse (Henriksen and Wesson 1978). The model possesses no unacceptably peculiar features apart (possibly) from its lack of homogeneity.

The cosmology described by equation (7.5) is, in fact, of the type proposed by De Vaucouleurs (1971) and others, in which small clumps are enclosed in larger clumps, the process being assumed to continue indefinitely and being therefore termed hierarchical. Matter in clusters and superclusters (that is, up to a scale of about 30 Mpc) is observed to be clumped inhomogeneously in this way, with a density profile around a local centre of inhomogeneity of the form $\rho \propto r^{-\Theta}$ with $\Theta \simeq 1 \cdot 7$. If this structure carries on indefinitely, the Universe might be hierarchical on all scales, possibly being described by a model such as equation (7.5). Such a model would seem to be superior to that of Bonner (1972) and an earlier one of Wesson, in that it possesses the cosmologically desirable property of being scale-free (cf Sciama 1971, p116, who, however, uses the term self-similarity to mean scale-free in a mathematical rather than physical sense). Many believe, of course, that the inhomogeneity observed on scales up to 30 Mpc dies out over greater distances and that the Universe overall is largely isotropic and homogeneous (see, for example, Peebles 1971). This may well be the case, but an interesting fact may be mentioned here in the opposite sense. The Einstein equations (7.3) are, on close examination, very easy to solve for inhomogeneous matter distributions with profiles of the sort $\rho \propto r^{-\Theta}$ with $\Theta = 2$. This seems to suggest that there is something fundamental about inverse-square density profiles, and the value $\Theta = 2$ is not far away from the observed clumped-galaxy value of $\Theta = 1\cdot7$. There is in fact an exact (static) solution to equations (7.3) which has $p \propto r^{-2}$ and $\rho \propto r^{-2}$ with an isothermal equation of state, and is perhaps the simplest solution one can find to Einstein's equations in the noted form.

The propensity for inverse-square density profiles to turn up both in solutions to Einstein's equations and in observed astrophysical systems (such as rich clusters of galaxies, which often follow the De Vaucouleurs rule with $\Theta \simeq 2$) cannot be an accident. Since the self-similar argument gives a solution with $\rho \propto R^{-2}$ exactly when the expansion velocity is small, one might suspect that this type of law is connected with an absence of scales. Further investigation (Henriksen and Wesson 1978) shows that this is so, and yields insight also into other types of scale-free configurations, including rotating ones. It is possible to combine results for both types of system and express the existence of scale-free matter and rotation laws in terms of two dimensionless numbers which appear to have considerable significance since the rules involved are observed in practice to apply in many astrophysical problems. The relations in question are $\rho_1 b_1{}^2 = \rho_2 b_2{}^2$, and $b_1 \Omega_1 = b_2 \Omega_2$ for two concentric systems of sizes b_1 and b_2, of mean densities ρ_1 and ρ_2 and angular velocities Ω_1 and Ω_2. The absolute numerical values concerned can be obtained from observation, and, coupled with suitable parameters, theory and observation for any astrophysical system can be summed up in the two dimensionless numbers

$$\eta_1 \equiv \frac{G\rho_s b_s{}^2}{c^2} \approx 10^{-7}; \qquad \eta_2 \equiv \frac{b_s \Omega_s}{c} \approx 10^{-4}. \qquad (7.6)$$

These dimensionless numbers seem to be fundamental. The forms of the relationships are confirmed on all scales from superclusters down to planets (Brosche 1974), where a prevalence for the angular momentum L_s of a system with moment of inertia I_s to be proportional to the square of the mass of the system M_s, has been known for a long time. Such a dependency is consistent with the relations given above because $M_s \propto \rho_s b_s^3$, so that $L_s = I_s \Omega_s \propto M_s^2$ becomes $\rho_s b_s^5 \Omega_s^2 b_s^6$ or $\rho_s b_s / \Omega_s =$ constant, as stated. (The $L_s \propto M_s^2$ rule is not completely free of ambiguity, however, as discussed by Hughes 1975b.) Further, the Regge trajectories on which elementary particles fall are characterised by a number L_s/M_s^2 (Regge) which differs by a large dimensionless number from the value L_s/M_s^2 (astrophysics). The form of these relationships, as pointed out by Brosche, would be difficult to explain if the fundamental parameters of physics were variable.

The two numbers in equation (7.5) are dimensionless and appear to have some basic significance in astrophysical matter distributions. Unlike the comments of §7.2, the results of this section are derivable using Einstein's theory of general relativity in its usual form. These results, however, are probably also consistent with a wider-ranging theory which includes general relativity as a special case. It is difficult to avoid this implication when one considers the coincidences noted previously, and one should logically consider such results derived within the confines of Einstein's theory as indications of how that theory might most sensibly be extended or modified. This topic will be returned to in §§7.5 and 7.6 after a brief comment on how the expansion argument as considered so far fits in with related processes in geophysics.

7.4. Geophysical Expansion

The result $\dot{R}_E = HR_E$ of §7.2 explains exactly, as noted above, the coincidence in the observed expansion velocity and that predicted by Hubble's law as discovered by MacDougall *et al* (1963; §6.5) and as quantitively supported by the data of §6.3. It has been apparent for some time that expansion as a geophysical process is a phenomenon that should be incorporated into geophysical theories (cf Chapter 5), and the available evidence would seem to be best explained by a combination of expansion and shallow convection.

The results of Creer (1965a, b) point strongly to expansion, as do the palaeogeographic data and the evidence of Chapter 6.

Convection in the account of Wesson (1972b) is confined to the region between continents, exists only down to 700 km depth, is irregular and generally very restricted. This is in agreement with the experimental result, used by Richter and Parsons and others (see Smith 1975b), that convective rolls are only stable for Rayleigh numbers less than about 13 times the critical one, above which a secondary set of cells develops at right angles to the first set. At more than 100 times the critical Rayleigh number the regime changes again to a pattern (in plan) resembling spokes radiating from the centre of a wheel. Below 700 km the model has a zone where the modified Lomnitz law of Jeffreys (1970, p11) is obeyed, while above that depth elastico-viscosity holds. The continents are gradually shifted along the direction of the radius vector at any point, by global expansion. This causes the great-circle distance between given points in any two continents to increase with time, but if the sialic masses do not grow or shrink, the ocean floor increases in surface area in between the continents, as discussed in §6.2. This relative motion is the basic driving force seen in sea-floor spreading, but is irregular in its manifestation on the Earth because of the great inhomogeneity of the lithosphere, although the total rate of spreading over long time intervals is not irregular since there is evidence that the rate of change of the planet's radius is constant. The dynamics of this process are coupled with unsteady convection as outlined above, since the region of the mantle above 700 km depth is on the verge of instability to mass motions generated by Nabarro–Herring creep (Wesson 1972b). This account, which is considered further in a volume on plate tectonics (Wesson 1974b, c), is clearly capable of explaining such data as those on mobile belts and the ophiolitic suites in such orogenic zones, usually advo-cated as support for whole-mantle convection. Along the lines of this model, sea-floor spreading is inevitable, but it is not certain, as sometimes discussed, that oceanic ridges are governed by local expansion and phase changes, since there is no evidence for phase changes at depths shallower than about 225 km, and the morphology and mineralogy of ridges (Wesson 1972a) give no indication of their extension to such depths, 4 km being nearer the mark (Cann 1970). There is no immedi-ate mechanism on this model for large-scale continental drift, but if such a mechanism is demanded while needing to avoid the implication of convection currents throughout the whole mantle, one could adopt a self-propulsive mechanism such as that of Elder (1967), or the interesting model of Pollack (1969) in which gravitational attraction between sea floor and sial causes the ocean floor to diverge at the ridges.

It would seem quite feasible to develop a detailed geophysical model in which expansion is the main driving force behind all large-scale geodynamic phenomena. Such a global process could provide energy for convection and orogenesis by a partial regulation of the geothermal flux and changes in the large-scale morphology of the planet. Sea-floor spreading, as noted, would be automatically explicable, and a firmer foundation for the hypothesis of plate tectonics than the familiar ones would thereby be provided.

7.5. The Significance of Λ

The type of model considered in §7.4 is an example within general relativity of a way in which gravitational theory might be developed to encompass some of the data considered in previous chapters. The cautious philosophy of extending known theory gradually, aided by comparison with evidence like that noted in Chapters 4 and 6, would seem to be more likely to yield a practicable wider theory than departures from new and untried principles. The latter type of development, as seen in Chapter 3, has a habit of going wrong.

While models of the self-similar type, such as equations (7.5), may be valuable as indicators of where new developments might be profitably sought, they fail in one important aspect, in that they have no term corresponding to the cosmological constant Λ of Einstein's theory. The reason is fundamental: Λ in general relativity has the dimensions of time^{-2} (or length^{-2}, the transition between the two being merely a factor c^2) and so $c\Lambda^{-1/2}$ is a unit of length which was ruled out at stage one by a wish to find a scale-free cosmology. Thus, the simplest type of self-similar cosmology fails to provide an opportunity to incorporate a Λ term into the model. (There is a more complicated type of self-similarity, discussed by Barenblatt and Zeldovitch

(1972) in which scales are present, but this seems undesirable for cosmological purposes.) What can be done about this?

One answer is, of course, simply to deny that Λ should be present at all. This used to be a popular opinion, but has gradually fallen into less favour because it is now widely realised that the real significance of Λ is that it is a way of incorporating the quantum physics of the vacuum into gravitational theory. The Λ term is also the obvious way to express the gauge nature of the space–time background within a theory of large-scale phenomena. It cannot be given up without losing the hope of incorporating elementary particle physics and gravitational physics into one formalism (see §2.8 for a discussion of this). That Λ represents a measure of the energy density and pressure of the vacuum is clear even in Einstein's theory.

It is a convenient fact that equation (7.1), with $\Lambda \neq 0$, matter pressure p_m, and matter energy density ϵ_m, is equivalent mathematically to equation (7.1) with $\Lambda \equiv 0$ and p replaced by $p_m + p_v$, ϵ by $\epsilon_m + \epsilon_v$ (Petrosian 1974), where

$$\epsilon_v = \frac{c^2 \Lambda}{8 \pi G} = -p_v. \tag{7.7}$$

This interpretation of Λ in terms of a constant background energy density and pressure has been known since Eddington's time, and is a simple way of including the properties of the vacuum in Einstein's equations. Solutions of equation (7.1) with $\Lambda = 0$, $p = p_m + p_v$ and $\epsilon = \epsilon_m + \epsilon_v$ are equivalent to solutions of equation (7.1) with $\Lambda \neq 0$, $p = p_m$ and $\epsilon = \epsilon_m$.

The finiteness of Λ is a question of dispute which, over the years, has become tedious through repetition. Essentially, equation (7.1) represents different sets of field equations (with different solutions) depending on whether Λ is retained or not (McCrea 1971). If Λ is kept finite, the de Sitter cosmology represents the natural end point of a series of matter-containing models as $\rho_m \to 0$. The influence of Λ on cosmology depends on the contending influences of ρ_m and ρ_v and this way of looking at Λ fits in well with the previously given discussion of Van Flandern's S hypothesis.

Both because one wishes to keep open the possibility of incorporating quantum and gravitational physics into one theory, and because it is easy then to understand the meaning of empty models as the limit of a class of matter-plus-vacuum solutions, one would like to keep Λ finite. In addition to these two reasons for including a Λ term in cosmology as a matter of logicality, the discussion of §2.8 showed that if the successful Salam–Ward–Weinberg type of elementary particle theory is indeed valid as an explanation of quantum phenomena, then Λ must be finite. Further, spontaneous symmetry-breaking theories suggest that Λ must not only be finite but may also be variable. The latter type of theory takes one outside the scope of Einstein's general relativity, and opens the field for a general discussion of how large a Λ term in the wide sense might be, and what evidence there is for its influence in astrophysics.

In general, models with notable Λ terms might be best treated using a bimetric approach (see, for example, Nelson 1972) in which the g_{ij} are expressed as sums of two parts, one representing the uniform background and the other representing the clumped matter. This approach allows one to use the background metric to provide a reference stratum to which the Cosmological Principle and other cosmological symmetry properties apparently refer. The Λ vacuum density can then be incorporated into this background if, as in Einstein's theory, it is homogeneous. This agrees with Horák (1970), who realised that the finiteness of Λ is crucial to an understanding of Mach's Principle and considered a model in which the Seeliger gravitational potential did not depend on location. A similar viewpoint has also been taken by Klein (1958, 1972) who has pointed out that Mach's Principle is only compatible with the Principle of Equivalence in a hierarchical-type Universe in which all systems should be expanding if the Cosmological Principle is to hold.

For models that do not depart too far from those specified by general relativity, the best limit on Λ comes from cosmological considerations (Petrosian 1974, p45) and is $|\Lambda| \lesssim 2 \times 10^{-56}$ cm^{-2}. A value of $|\Lambda| \simeq 1 \times 10^{-56}$ cm^{-2} is equivalent, by equation (7.7), to a density $\rho_{\rm v} = |\epsilon_{\rm v}|/c^2 \simeq 10^{-29}$ g cm^{-3}, compared with the observed matter density of $\rho_{\rm m} \simeq 10^{-30.5}$. The critical density for the Universe, $\rho_{\rm c} = 3H^2/8\pi G$, is also $\rho_{\rm c} \simeq 10^{-29}$ g cm^{-3} if $H \simeq 2 \cdot 5 \times 10^{-18}$ s^{-1} ($\equiv 75$ km s^{-1} Mpc^{-1}). The dynamical effect of the uniform $\rho_{\rm v}$ can therefore dominate.

208

This result gains support from dynamical studies of the velocity field of local galaxies. Sandage *et al* (1972) found that the observed local clustering of galaxies did not, surprisingly, affect departures from a uniform flow, as expected. This phenomenon has been much discussed (for example, Tamman 1974) and means either (*a*) that the kinetic term T is very large $(T \gg |W|)$ in the virial expression $T + W = E$, or (*b*) there exists an unseen background of uniformly distributed density $\langle \rho \rangle \gg 10^{-30 \cdot 5}$ g cm^{-3} as seen in the galaxies. Explanation (*b*) is, in a sense, a prediction of any model having a large background energy density ϵ_V provided by a Λ term.

This does not necessarily prove that $\Lambda \neq 0$, since the work of Sandage *et al* (1972) has been shown to be partly incorrect as a critique of inhomogeneous world models. De Vaucouleurs has remarked that the density law used is not conceptually the same as that of De Vaucouleurs (1971), while Rowan-Robinson (1972) has pointed out an error in the counting of faint, remote galaxies which, when corrected, tends somewhat to uphold the presence of inhomogeneity. The data of Sandage *et al* (1972) have been used by Le Denmat *et al* (1975) in confirming an apparent anisotropy in Hubble's parameter discovered earlier by Rubin *et al* (1973) using observations of ScI-type galaxies. This anisotropy, if it really is an anisotropy in H, seriously compromises homogeneous cosmological models. While it is possible to explain at least part of the anomaly on a two-cloud model of the clumpy distribution of local galaxies (Fall and Jones 1976), it would seem quite likely that the most natural explanation of it is in terms of a peculiar velocity of the Galaxy and the Local Group with respect to more remote galaxies. The very detailed account that reaches this conclusion (Rubin *et al* 1976a, b) puts the velocity at 454 ($\pm$ 125) km s^{-1} in the direction towards Galactic coordinates $l = 163°$, $b = -11°$. This interpretation is potentially of enormous significance, since it is equivalent to a detection of the Galaxy's peculiar velocity with respect to the 3 K microwave background, and raises the possibility (discussed below) that the background is not the usually assumed remnant of a big bang fireball. The only really satisfactory way to avoid this conclusion is to accept the equally upsetting one that H is genuinely anisotropic. The regional discrepancy of about 20% of the mean value is then significant, although the way in which H depends on direction and distance is still not closely defined. The result of Le Denmat *et al* (1975) is that H ($\lesssim 30$ Mpc distances) $\simeq 95$ (± 11) km s^{-1} Mpc^{-1} while H (> 30 Mpc) = 63 (± 5) km s^{-1} Mpc^{-1}.

A similar confirmation has been found by Guthrie (1976), who has extended the region over which the anomaly can be detected to one in which recession velocities are in the range $4000 < v_r < 25\,000$ km s^{-1}, using data on the brightest galaxies in clusters. These results have relevance to the size of the Λ term, since they give information on the large-scale dynamics of the Universe. The work of Le Denmat *et al* incorporates the effect of De Vaucouleurs' Local Supercluster, which, despite the criticisms of Jones (1976) is now realised to be a real entity (he considered the Local Supercluster to be an artefact produced by a large hole in the anticentre region, that is, away from the Virgo cluster, but covariance analysis leaves no doubt that superclusters exist). Indeed, even the Coma cluster has a size of $\gtrsim 30$ Mpc (Chincarini and Rood 1975), as revealed by delicate photometry, and represents a massive ($\gtrsim 1 \times 10^{16} M_\odot$) inhomogeneity in which crossing times, if definable at all, are at least 4×10^{10} yr. The existence of the Local Supercluster, in which the Local Group is embedded (De Vaucouleurs 1975a) among intergalactic gas clouds (De Vaucouleurs and Corwin 1975, De Vaucouleurs 1975b, c), is therefore representative of an expected group of phenomena.

The existence of such matter inhomogeneities as superclusters and possible anisotropies in Hubble's parameter H raises doubts about the cosmological origin of the microwave background. This question has been examined elsewhere (Wesson 1975a). There are several analyses (for example, Dautcourt 1969, Dyer 1975) of the expected effects of inhomogeneities in the matter distribution on the 3 K field, and there is some evidence from cosmic ray studies that the universality of the microwave background may need modifying if the expected cut-off in cosmic ray particle energies at about 5×10^{19} eV, due to their scattering off the 3 K photons to produce photo-mesons, continues to go unobserved. The absence of the expected cut-off up to $\simeq 1 \times 10^{20}$ eV has led to the suggestion that the ultra-high-energy cosmic rays originate

210

somewhere in the halo of the Galaxy or the Supercluster (Wdowczyk and Wolfendale 1975). A non-remote source is supported by a study of the size distribution of cosmic γ ray bursts (Bewick et al 1975), which tend to imply a Galactic origin for the particles producing the smaller bursts, if not the larger, while it seems possible to confine some high-energy cosmic rays ($\approx 10^{12}$ eV) within the Galaxy by scattering off Alfvén waves produced by the low-energy component (Skilling 1975). Alternatively, Scott (1975) has studied the possibility that cosmic rays propagate to the Earth along low-density tunnels in the interstellar medium, having their origin in local supernovae events, or else populate a low-density halo. While cosmic ray data have yet to raise any serious doubt as to the cosmological origin for the 3 K field, as opposed to a possible discrete-source origin (Wolfe and Burbidge 1969), or an origin from thermal dust emission (Narlikar and Wickramasinghe 1968, Wesson 1974a), or dipole dust emission (Wesson and Lermann 1976), one interesting fact should be pointed out. An anisotropy in the microwave field *was* detected by Partridge and Wilkinson (1967). It was detected also by Conklin (1969) and by Henry (1971). Both of the last two references agree in putting the anisotropy, which is of size 3·2 ($\pm 0·8$) mK (corresponding to a peculiar velocity of about 300 km s^{-1}) in the direction 10 h (RA), -30 degrees (dec). Thus there is positive evidence for a persistent anisotropy of the microwave field, although it is usually removed by a process of correction to some chosen standard of rest (see Partridge 1974 for a review). The latter procedure seems bafflingly tautologous when examined in detail (Wesson 1975a), and Fox et al (1975), following an earlier incorrect formulation by Von Hoerner (1973), have shown that it is impossible to define an objective concept of rest in Robertson–Walker cosmologies.

If the microwave background is anisotropic to any degree, it will upset the universality of the photon/baryon ratio mentioned in Chapter 4. This ratio is commonly believed to have been fixed near the origin of the Universe, perhaps, as in the hypothesis of Dehner and Hönl (1975a, b) by the mutual annihilation of nucleons and antinucleons in the big bang, leaving only 10^{-7} of the original baryons and resulting in a photon/baryon number of the order of 10^8. Davies (1974c, p196) has shown that a photon/baryon number of $\approx 10^{10}$ can be generated quite naturally by the burning of protons inside stars over the history of the Universe with the release of 0·01 of the rest mass as starlight. He has also considered a closed-time oscillating Robertson–Walker model of the conventional ($k = +1$) type, in which the starlight is thermalised at the bounce and appears as the 3 K background in the next cycle. It is also possible, in view of results on entropy production from exploding black holes (Davies and Taylor 1974, Kundt 1976), that primaeval singularities may have produced the microwave background. The latter hypothesis seems very likely, since Parker (1976) has found that a Friedmann Universe, when still near the singular stage and when at the Planck temperature of 10^{32} K, can spontaneously create particles. If the particles are photons, they are produced with a black-body spectrum and an entropy compatible with the observed background. Alternatively, it may be that the long-wavelength photons of the 3 K background are radiation which has fallen into our (closed) Universe from outside, like light into a black hole, and been red-shifted from starlight to microwaves (Breuer and Ryan 1975). Some such mechanism as one of the four just mentioned seems to be required since Chapline (1976) has used a quark theory of hadrons to show that the entropy of the presently observed Universe could not have been produced by the dissipation of low-level fluctuations, as is often assumed. Chapline's results indicate that the early Universe may have been inhomogeneous or anisotropic, and raises the question of how closely it was Friedmannian at those and later times.

In connection with this, Gunn and Tinsley (1975) believe that the local Universe can be fairly well described by the Friedmann models, and have collected together data on the Hubble diagram, estimates of the observed matter density and other parameters which indicate that the Universe will probably continue to expand forever and that it is likely that Λ is large in magnitude, and positive.

7.6. A Need for a Wider Theory?

In order to explain some puzzling coincidences in geophysical and astronomical data, one can in principle assume that a metric theory of gravitation exists which, in some special gauge, gives equations that are formally the same as those of Einstein's theory. The latter, under specified assumptions, can be used to gain insight into certain processes which otherwise seem unconnected (§§7.2 and 7.4). In particular, one can avoid invoking G-variability by using Van Flandern's S hypothesis instead. While this may be desirable from the standpoint of departing as little as possible from established theory, it introduces certain unfamiliar properties connected with metric continuity and scale transformations. A consideration of the latter naturally leads to the concept of scale-free cosmology (§7.3), a concept that is compatible with general relativity but probably also indicates a direction in which gravitational theory can be rewardingly extended. Scale-free cosmology, within the confines of Einstein's theory, has no natural place for a cosmological term Λ, even though there is considerable evidence that Λ is finite and, cosmologically speaking, of large size (§7.5). Quantum theory presents the possibility of a variable Λ. Using the increased scope that this possibility offers, one is able to speculate on the chances of combining the indications of previous sections to obtain a pointer to a wider theory that might include explanations of previously noted data in a more or less direct manner.

One way to obtain a self-similar cosmology without scales and yet include a Λ term, is to make Λ variable by writing $\Lambda = \Lambda_0/t^2$, where Λ_0 is a dimensionless constant. There is now no problem with the introduction of an unwanted length scale. This innovation also introduces a cosmological term in a way that is aesthetically more satisfying than that of the Einstein theory. In general relativistic cosmological models with a Λ term, the model expands with the matter density systematically decreasing with time, but with the equivalent Λ density ρ_v and Λ pressure p_v remaining constant with values given by equation (7.7). If one introduces a positive matter pressure, perhaps due to intergalactic cosmic rays (Blair 1974) or to the 3 K microwave background (Sciama 1971), equation (7.3.2) shows that all systems lose mass, but the vacuum background remains unaffected. This inviolate aspect of the model's substratum seems somehow unsatisfactory. A more reasonable picture might be one in which the matter properties vary in unison with the vacuum properties, and where the direction of the variations depend on the sign of Λ and any matter pressure present in the model. This would be one way to achieve a kind of balance between the matter and the background.

Another is suggested by the fact that most astrophysical systems, whether stars, galaxies or clusters of galaxies, appear to possess a positive, internal kinetic pressure that must, by equation (7.3.2) lead to mass loss unless these systems are isolated in an absolutely perfect, pressure-free space. Such mass loss might be compensated by the creation of new atoms in the interiors of these objects, perhaps in the manner envisaged by one of the theories discussed in Chapter 2. This hypothesis depends, however, on the use of an equation such as (7.3.2). There is no guarantee that such an equation will hold in an extended theory, but if equations (7.3) do retain their validity in one gauge of a more comprehensive theory, then general relativity may be like a hydrodynamical model in which the equations are formally correct only because sources (and maybe sinks) of matter exist that, while not explicitly included in the theory, allow the existence of the flows prescribed by the equations.

While the two above suggestions are to a large extent *ad hoc*, they are not mutually exclusive and do indicate plausible ways in which the existing theory could be extended. The implication, therefore, is that there could well be a genuine requirement in cosmology for a theory, such as the complete Dirac ($\times$) model of §2.3, in which there is continuous creation of matter where matter is already most dense, and in which a Λ term could be of major importance in explaining the large-scale structure and evolution of the Universe.

8. Summary

With the aid of hindsight, it can be seen that cosmology as it relates to geophysics is a subject that has provided, over the history of the systematic study of the two subjects, many new ideas and suggestions, both for these two disciplines, and also for numerous attendant fields of study. Original hypotheses in both main subjects have become progressively more sophisticated, with the gradual emergence of certain clear and persistent lines of thought that can now be readily discerned. It is the purpose of this chapter to summarise the main conclusions of previous pages, and to scrutinise carefully the most acceptable hypotheses with a view to seeing how they interrelate.

Cosmologies which have usually been thought of as possessing the most direct relevance to geophysics have been those in which the Newtonian gravitational parameter G is time-variable. There are also other formalisms that lack adequate theoretical foundations, but which may turn out to be pertinent geophysically. The account of Chapter 2 showed that the most notable theories to survive a preliminary comparison with observational data are those of Brans and Dicke, Hoyle and Narlikar, and the complete Dirac theory. Of these, the first has difficulties with solar oblateness data, but is not seriously in conflict with other material. The second has difficulties with some astrophysical data and limits set on $|\dot{G}/G|$, but has a very strong logical basis in conformal invariance and the Wheeler–Feynman absorber theory. The third does not conflict seriously with any extant data, but has a less satisfactory intuitive foundation than the other two because of the obscurity surrounding the meaning of the large dimensionless cosmological numbers and their coincidences. In Chapter 3 it was seen that there exists no adequate account of the origin of inertia, which should surely be a prerequisite for the eventual

215

acceptance of any new theory of gravity of the variable-G type. However, various pieces of evidence suggest that inertia produces little anisotropy locally, and is a $1/r$ effect. There is no evidence that G is noticeably space-variable, and the absolute value of G may be connected with properties of the vacuum. The latter need to be approached via particle physics and not gravitational theory. Metric theories of gravity have been intensively studied, the only surviving ones being general relativity, the Brans–Dicke cosmology and the vector–tensor theory. One must also admit to this select group, on grounds of relatedness, the Hoyle–Narlikar and complete Dirac theories.

One important point emerges about all the theories which survive empirical and theoretical examination. The meanings of coordinate and reference frames have gradually been better understood in subjective, rather than mathematical, terms, although this does not imply any loss of objectivity or rigour. The essential arbitrariness was recognised long ago by Milne, who developed a theory of kinematic equivalence in which only clocks were fundamental. It is most powerfully used in the Hoyle–Narlikar cosmology, and in other places where conformal invariance is demanded.

Chapter 4 reviewed the subject of the Eddington numbers and the cosmological coincidences, as these numbers are ubiquitous in the $\dot{G}$ cosmologies and in other astrophysical connections. While there is a tendency to believe that these numbers are connected with elementary particle physics in some way, and may be predicated on our existence as observers, there does not exist any readily acceptable account of them, despite much competent work that has been done. There is no positive evidence that any of the physical parameters, other than G, are variable in space or time. Chapter 5 demonstrated that, despite the presence of numerous complex perturbations, the rotation of the Earth provides the longest-running reasonably accurate clock available to cosmology. The secular spin-down may, or may not, contain a cosmological term. The data are compatible with G decreasing at a rate $|\dot{G}/G| \approx 10^{-11}$ yr^{-1}, or with large global expansion. A finite measurement of $\dot{G}/G = -8\ (\pm 5) \times 10^{-11}$ yr^{-1} could be re-interpreted as expansion, or secular mass loss.

Chapter 6 examined the expanding Earth hypothesis, which

is a recurrent topic in geophysics. It has two aspects, the first being the moderate amounts of expansion predicted by the acceptable $\dot{G}$ cosmologies ($\simeq 500$ km), the other being the suggestion that the globe has expanded steadily over its history at a rate of about $0{\cdot}5$ mm yr^{-1}, following some cosmological process (total expansion $\simeq 3000$ km). There is no ready evidence to disprove the former alternative. The latter hypothesis has some evidence in favour of it and some against, but the evidence is very difficult to judge because most of the expansion would have occurred long ago in the Pre-Cambrian. Some tests of the hypothesis are, however, possible. Chapter 7 explored the possibility of avoiding G variability by adopting in its place Van Flandern's S (expansion) hypothesis. Self-similarity seems to be the natural way to develop research in cosmology if the latter is to be free of scales, but a new theory is probably needed if one is to heed the strong evidence indicating that a (possibly variable) cosmological Λ term is needed in cosmology.

The main tasks of experimental physics in refining limits on acceptable parameter variabilities in metric theories and in attempting to confirm or refute geophysical hypotheses such as that of expansion (Jordan 1973) are certainly not less difficult than the theoretical problems involved. Of the latter, it would appear that conceptual problems of a very fundamental type represent the greatest tasks for theoretical physics as it applies to cosmology and geophysics. Two major problems are apparent: firstly, the meaning of coordinate parametrisation and the use of certain preferred frames in relativistic theories is a basic problem that is intimately bound up with clock graduation and ideas on the meaning of entropy; the other problem concerns the nature of inertia. Both topics have seen the expenditure of enormous amounts of effort, and remain relatively little understood.

One of the few concrete results in the parametrisation problem concerns conformal invariance and the use to which it has been put in filtering out acceptable cosmological models in the Hoyle–Narlikar cosmology. This theory is also notable in that it makes explicit use of the Wheeler–Feynman absorber electrodynamics, and attempts a rational interpretation of the theory of entropy which ensues. The classic subject of entropy and the direction of time's arrow has a history as long as that of cosmology, and is somewhat more inconsequential. Davies (1974c) has reviewed the whole topic of time asymmetry. As mentioned previously, it is probably true that the familiar increasing tendency of entropy with time is a result of the continual formation of branch systems. This local phenomenon does not seem to solve the cosmological problem, but it may have a very simple explanation. Milne realised that to define the increase in entropy of a system requires the axiom that it is possible to divide the system into two parts, such that one part is unaffected by the entropy-producing process and can be compared with the other part (see Whitrow 1961, p7). The Universe, as the term is often understood, cannot be divided into two parts in this way if it is infinite. If it is finite, with what do we compare the entropy of our Universe? Another one? The possibility of there being many universes is a topic of occasional discussion in cosmology (Sciama 1973 has discussed the uniqueness of the Universe), but any sensible definition of the Universe would have to be an implicit one. Only hierarchical cosmology, with its many problematic observational aspects, can provide such a definition immediately. The evolution of entropy would certainly seem to be connected with our sense of time (Uhlenbeck 1973, Prigogine 1973). This suggests that careful consideration should be given to the exact interpretation to be put on time as it enters into the theories discussed in Chapter 2 (Hoyle 1966, Eigen 1973, Ellis 1974). One difficulty for the further development of these cosmologies concerns the prevalence, noted in Chapter 4, of theories which leave no room for extensions to include yet-undiscovered data. The need for a theory with a certain flexibility of this type is apparent in the search for explanations of the cosmological coincidences.

The Eddington numbers, besides being of distinctive sizes, have the rather baffling property of turning up in all kinds of calculations from the scale of the whole Universe, down to the scale of the elementary particle. This would seem to reinforce the opinion that any theory aiming at their explanation ought to be an open-ended one, since a definitive, closed theory of phenomena encompassing the entire range would be unnecessarily limited. (Attempts to construct unified field theories of

218

the strong, electromagnetic, weak and gravitational forces are, of course, still being made. See, for example, Llewellyn-Smith 1975 for gauge theories.) The flexibility requirement, besides being historically justified, might also suggest that the problem of the Eddington numbers is not, in principle, solvable with conventional mathematical approaches. Ulam (1972) in discussing Gamow's attempts to explain the numbers, mentions that both workers have considered it likely that even so basic a question as the finiteness, or otherwise, of the Universe may be undecidable. Gamow's belief in the local validity of Friedmann cosmology with $k = -1$ would mean that space–time, being non-compact, would necessarily contain a number of stars or galaxies that is denumerably infinite. Under these conditions, it is difficult to see how any physical theory, even one involving the Eddington number of the order of 10^{80}, could hope to be complete. Indeed, the system of mathematics used at the familiar level is amply rich enough to show that, fundamentally, not all the propositions it is possible to ask can be answered (Gödell 1962; Von Weizächer 1973 discusses logic in relation to physics). While it would obviously be difficult to decide which of the well known intractable cosmological problems might, in principle, be unsolvable, the need to extend the type of analysis employed on such questions gives collateral support to the desirability of developing new cosmologies. The reason is that departures such as that of Hoyle and Narlikar require the introduction of new mathematical approaches which increase the scope of the problems that can be tackled profitably.

The questions of the cosmological coincidences and the meaning of entropy might be ones that could benefit from the application of a more flexible approach to developing a workable formalism, yet explicit attempts to construct theories of physics that are self-consistent, but 'open', are not common. The Hoyle–Narlikar cosmology makes some attempt to do so, but the only really outstanding example is Hagedorn's statistical thermodynamics of strong interactions (§4.7). In this theory, hadrons are composed of each other, and the number of different kinds of them, obtained by integrating the mass spectrum, is infinite. The way in which the theory is developed has connections with a type of procedural paradox that is sometimes met in physics when quantities have to be defined in terms of themselves (Von Weizächer 1973, Fraenkel and Bar-Hillel 1958). From this point of view, it will be interesting to see if the Hagedorn theory is fully compatible with experimental data and continues to be in agreement with astrophysical results on conditions in the early Universe.

The foregoing comments suggest that fundamental progress in cosmology, in the wide sense of the word, is likely to be made through a clarification of the concepts used, rather than in refining numerical data. The activity in cosmology which has been in progress since the formulation of general relativity has largely been inspired by a wish to find some connection between physics on the large scale and physics on the small scale. Variable-G theories and searches for (possibly empty) meanings of the Eddington numbers are both symptoms of this. The fact that there is still no definite evidence of any of the effects expected from these theories is not simply an indication of a desire to engage in speculation. It is rather an urge to find a hypothesis that connects a large number of observed coincidences in various regions of astronomy and geophysics. The persistence of research in the areas discussed in the face of little positive return is evidence of the conviction felt among the members of a large class of workers that some such hypothesis does indeed exist. Attractive as it may be to pursue unified theories, history shows that any such hypothesis that might be discovered would probably be open-ended in the sense discussed above. The search for a theory that encompasses data outlined in previous chapters, but does not constrain future developments, would certainly seem to be justifiable, even if it is becoming ever more apparent that the task is not an easy one.

References

Adams P J, Hsieh S H and Tsiang E 1976 *Nature* **264** 485

Alfvén H 1964a *Icarus* **3** 52–6

—— 1964b *Icarus* **3** 57–62

Alley C O, Bender P L, Currie D G, Dicke R H and Faller J E 1969 *The Application of Modern Physics to the Earth and Planetary Interiors* ed S K Runcorn (New York: Wiley) pp523–30

Alpher R A and Gamow G 1968 *Proc. Natn. Acad. Sci. USA* **61** 363–6

Alpher R A and Herman R 1972 *Cosmology, Fusion and Other Matters* ed F Reines (Bristol: Adam Hilger) pp1–14

Ambartsumian V A 1958 *La Structure et l'Evolution de l'Univers* (Brussels: Stoops) pp241–74

Anderle R J 1972 *Rotation of the Earth* (*IAU Symp. No.* 48) ed P Melchior and S Yumi (Dordrecht, Holland: Reidel) pp101–3

Anderson D L 1974 *Science* **182** 49–50

Anderson D L, Sammis C and Jordan T 1972 *The Nature of the Solid Earth* ed E C Robertson (New York: McGraw-Hill) pp41–66

Anderson J L and Finkelstein D 1971 *Am. J. Phys.* **39** 901–4

Aoki S 1967 *Publ. Astron. Soc. Japan* **19** 585–95

Armstrong R L 1969 *Nature* **221** 1042–3

Arnowitt R, Deser S and Misner C W 1960 *Phys. Rev.* **120** 313–20

Asnani H 1969 *Nature* **222** 968–9

Aspden H 1966 *The Theory of Gravitation* 2nd edn (Southampton: Sabberton)

Aspden H and Eagles D M 1972 *Phys. Lett.* **41A** 423–4

Audouze J 1971 *Nature* **231** 381

Bahcall J N and Salpeter E E 1965 *Astrophys. J.* **142** 1677–81

Bahcall J N, Sargent W L W and Schmidt M 1967 *Astrophys. J. Lett.* **149** L11–5

Bahcall J N and Schmidt M 1967 *Phys. Rev. Lett.* **19** 1294–5

Bahcall J N and Woltjer L 1974 *Nature* **247** 22–3

Bailey V A 1960 *Nature* **186** 508–10

Barbour J B 1974 *Nature* **249** 328–9

——1975 *Nuovo Cim.* **26B** 16–24

Barenblatt G D and Zeldovitch Y B 1972 *Ann. Rev. Fluid Mech.* **4** 285–312

Barnes A and Whitrow G J 1970 *Mon. Not. R. Astron. Soc.* **148** 193–5

Barnett C H 1962 *Nature* **195** 447–8

—— 1969 *Nature* **221** 1043–4

Barnothy J 1969 *Astrophys. J. Lett.* **156** L159–60

Barnothy J and Tinsley B M 1973 *Astrophys. J.* **182** 343–9
Baum W A and Florentin-Nielsen R 1976 *Astrophys. J.* **209** 319–29
Beams J W 1971 *Physics Today* **24** 35–40
Beck A E 1969 *The Application of Modern Physics to the Earth and Planetary Interiors* ed S K Runcorn (New York: Wiley) pp77–83
Begelman M C and Rees M J 1976 *Nature* **261** 298–9
Bender P L, Currie D G, Dicke R H, Eckhardt D H, Faller J E, Kaula W M, Mulholland J D, Plotkin H H, Poultney S K, Silverberg E C, Wilkinson D T, Williams J G and Alley C O 1973 *Science* **182** 229–38
Berry W B N and Barker R M 1968 *Nature* **217** 938–9
Bertotti B 1962 *Proc. 20 Course of Enrico Fermi International School of Physics* (New York: Academic Press) pp174–201
—— 1964 *Relativistic Theories of Gravitation* ed L Infeld (Oxford: Pergamon) pp91–2
Bertotti B, Brill D and Krotkov R 1962 *Gravitation: an Introduction to Current Research* ed L Witten (New York: Wiley) pp1–48
Bewick A, Coe M J, Mills J S and Quenby J J 1975 *Nature* **258** 686–7
Binder A B 1966 *Science* **152** 1053–5
Binder A B and McCarthy D W 1972 *Science* **176** 279–81
Birch F 1967 *Phys. Earth Planet. Interiors* **1** 141–7
Bishop N T and Landsberg P T 1976 *Nature* **264** 346–7
Blackett P M S 1947 *Nature* **159** 658–66
Blair A G 1974 *Confrontation of Cosmological Theories with Observational Data (IAU Symp. No. 63)* ed M Longair (Dordrecht, Holland: Reidel) pp143–55
Blin-Stoyle R J 1975 *Nature* **257** 179–80
Blow R A and Hamilton N 1975 *Nature* **257** 570–2
Bondi H 1952 *Cosmology* (London: Cambridge University Press)
Bondi H and Gold T 1948 *Mon. Not. R. Astron. Soc.* **108** 253–70
Bonner W B 1972 *Mon. Not. R. Astron. Soc.* **159** 261–8
—— 1974 *Mon. Not. R. Astron. Soc.* **167** 55–61
Bostrom R C, Sherif R C and Stockman R H 1974 *Plate Tectonics—Assessments and Reassessments* ed C F Kahle (Tulsa: American Association of Petroleum Geologists) pp463–85
Bottinelli L and Gougenheim L 1973 *Astron. Astrophys.* **26** 85–9
Braginskii V B and Ginzburg V L 1974 *Sov. Phys.–Dokl.* **19** 290–1
Brans C 1961 *PhD Thesis* Princeton University, NJ, USA
Brans C and Dicke R H 1961 *Phys. Rev.* **124** 925–35
Breuer R A and Ryan M P 1975 *Mon. Not. R. Astron. Soc.* **171** 209–18
Brill D R 1962 *Proc. 20 Course of Enrico Fermi International School of Physics* (New York: Academic Press) pp50–68
Brosche P 1974 *Astrophys. Space Sci.* **29** L7–8
Broulik B and Trefil J S 1971 *Nature* **232** 246–7
Cahill M E and Taub A H 1971 *Commun. Math. Phys.* **21** 1–40
Cann J R 1970 *Nature* **226** 928–30
Cannon W 1974 *Phys. Earth Planet. Interiors* **9** 83–90
Canuto V, Adams P J and Tsiang E 1976 *Nature* **261** 438

Carey S W 1961 *Nature* **190** 36–7

—— 1975 *Earth Sci. Rev.* **11** 105–43

Carter B 1974 *Confrontation of Cosmological Theories with Observational Data* (*IAU Symp. No.* 63) ed M Longair (Dordrecht, Holland: Reidel) pp291–8

Cathles L M 1975 *The Viscosity of the Earth's Mantle* (Princeton, NJ: Princeton University Press)

Cavallo C 1973 *Nature* **245** 313–4

Challinor R A 1971 *Science* **172** 1022–5

Chapline G F 1976 *Nature* **261** 550–1

Chapman D and Pollack H 1975 *Earth Planet. Sci. Lett.* **28** 23–8

Chapman S 1929 *Nature* **124** 19–26

Chin C W and Stothers R 1975 *Nature* **254** 206–7

Chincarini G and Rood H J 1975 *Nature* **257** 294–5

Chinnery M A and Wells F J 1972 *Rotation of the Earth* (*IAU Symp. No.* 48) ed P Melchior and S Yumi (Dordrecht, Holland: Reidel) pp215–20

Chiu H Y 1963 *Brandeis Summer Sch. Theor. Phys., Lectures on Astrophysics and Weak Interactions* vol 2 ed S Hayakawa *et al* (Englewood Cliffs, NJ: Prentice-Hall) pp165–275

Clegg P E, Rowan-Robinson M and Ade P A R 1976 *Astron. J.* **81** 399–406

Clutton-Brock M 1974a *Astrophys. Space Sci.* **30** Ll–3

—— 1974b *Astrophys. Space Sci.* **30** 395–408

Cocconi G and Salpeter E 1958 *Nuovo Cim.* **10** 646–51

—— 1960 *Phys. Rev. Lett.* **4** 176–7

Cohen E R and Du Mond J W M 1965 *Rev. Mod. Phys.* **37** 537–94

Cohen S A and King J G 1969 *Nature* **222** 1158–9

Colella R, Overhauser A W and Werner S A 1975 *Phys. Rev. Lett.* **34** 1472–4

Collins C B and Hawking S W 1973 *Astrophys. J.* **180** 317–35

Conklin E K 1969 *Nature* **222** 971–2

Countillet V E and Allegre C J 1975 *Earth Planet. Sci. Lett.* **25** 279–85

Cox A and Doell R R 1961 *Nature* **189** 45–7

Cox J P and Giuli R T 1968 *Principles of Stellar Structure* (London: Gordon and Breach)

Crain I K and Crain P L 1970 *Nature* **228** 39–41

Crain I K, Crain P L and Plant M G 1969 *Nature* **223** 283

Creber G T 1975 *Growth Rhythms and the History of the Earth's Rotation* ed G D Rosenberg and S K Runcorn (New York: Wiley) pp75–87

Creer K M 1965a *Nature* **205** 539–44

—— 1965b *Discovery* **26** 34–9

—— 1967 *International Dictionary of Geophysics* ed S K Runcorn (Oxford: Pergamon) pp382–9

—— 1975 *Growth Rhythms and the History of the Earth's Rotation* ed G D Rosenberg and S K Runcorn (New York: Wiley) pp293–318

Dan'us J 1976 *New Scientist* **72** 202–4

Darius J 1972 *New Scientist* **53** 482–3

Datt B 1938 *Z. Phys.* **108** 314–21

Dautcourt G 1969 *Mon. Not. R. Astron. Soc.* **144** 255–78

Davidson W and Narlikar J V 1966 *Rep. Prog. Phys.* **29** 539–622
Davies P C W 1972 *J. Phys. A: Gen. Phys.* **5** 1296–304
—— 1974a *Nature* **249** 208–9
—— 1974b *Nature* **249** 510–1
—— 1974c *The Physics of Time Asymmetry* (Guildford: Surrey University Press)
—— 1974d *Nature* **250** 460
—— 1975 *Nature* **255** 191–2
—— 1976 *Nature* **259** 157
Davies P C W and Taylor J G 1974 *Nature* **250** 37–8
Dearborn D S and Schramm D N 1974 *Nature* **247** 441–3
Dearnley R 1965 *Nature* **206** 1284–90
—— 1969 *The Application of Modern Physics to the Earth and Planetary Interiors* ed S K Runcorn (New York: Wiley) pp103–110
Dehner H and Hönl H 1969 *Astrophys. J. Lett.* **155** L35–42
—— 1975a *Astrophys. Space Sci.* **33** 49–57
—— 1975b *Astrophys. Space Sci.* **36** 473–8
Dehner H and Obregón O 1971 *Astrophys. Space Sci.* **14** 454–9
—— 1972 *Astrophys. Space Sci.* **15** 326–33
Dennis J G 1962 *Nature* **196** 364
Dennison B and Mansfield V N 1976 *Nature* **261** 32–4
De Vaucouleurs G 1971 *Publ. Astron. Soc. Pacific* **83** 113–43
—— 1972 *Nature* **236** 166
—— 1975a *Astrophys. J.* **202** 319–26
—— 1975b *Astrophys. J.* **202** 610–5
—— 1975c *Astrophys. J.* **202** 616–8
De Vaucouleurs G and Corwin H G 1975 *Astrophys. J.* **202** 327–34
Dicke R H 1953 *Phys. Rev.* **89** 472–3
—— 1957a *Rev. Mod. Phys.* **29** 355–62
—— 1957b *Rev. Mod. Phys.* **29** 363–76
—— 1958 *J. Wash. Acad. Sci.* **48** 213–23
—— 1959a *Nature* **183** 170–1
—— 1959b *Science* **129** 621–4
—— 1960a *Am. J. Phys.* **28** 344–7
—— 1960b *Quantum Electronics* ed C H Townes (New York: Columbia University Press) pp572–80
—— 1961a *Nature* **192** 440–1
—— 1961b *Phys. Rev. Lett.* **7** 359–60
—— 1962a *Science* **138** 653–64
—— 1962b *Proc. 20 Course of Enrico Fermi International School of Physics* (New York: Academic Press) pp1–49
—— 1962c *Rev. Mod. Phys.* **34** 110–22
—— 1962d *Phys. Rev.* **125** 2163–7
—— 1964a *Relativity, Groups and Topology* ed C and B De Witt (London: Gordon and Breach) pp165–313
—— 1964b *The Theoretical Significance of Experimental Relativity* (New York: Gordon and Breach)

Dicke R H 1966 *The Earth–Moon System* ed B G Marsden and A G W Cameron (New York: Plenum) pp98–164
—— 1968 *Astrophys. J.* **152** 1–24
—— 1969 *J. Geophys. Res.* **74** 5895–901
—— 1970 *Gravitation and the Universe* (Philadelphia: American Philosophical Society)
Dicke R H and Goldenberg H M 1974 *Astrophys. J. Suppl.* **27** 241–86
Dicke R H and Peebles P J E 1965 *Space Sci. Rev.* **4** 419–60
Dingle H 1954 *The Sources of Eddington's Philosophy* (London: Cambridge University Press)
Dirac P A M 1937 *Nature* **139** 323
—— 1938 *Proc. R. Soc.* **A165** 199–208
—— 1973a *Pont. Acad. Comment.* **11** 1–7
—— 1973b *Proc. R. Soc.* **A333** 403–18
—— 1973c *The Physicist's Conception of Nature* ed J Mehra (Dordrecht, Holland: Reidel) pp45–59
—— 1974 *Proc. R. Soc.* **A338** 439–46
—— 1975 *Nature* **254** 273
Disney M J, McNally D and Wright A E 1969 *Mon. Not. R. Astron. Soc.* **146** 123–60
Dodson M H 1971 *Nature* **234** 212
Doell R R and Cox A 1972 *The Nature of the Solid Earth* ed E C Robertson (New York: McGraw-Hill) pp245–84
Domokos G, Janson M M and Kovesi-Domokos S 1975 *Nature* **257** 203–5
Dooley J C 1973 *Search* **4** 9–15
Dubourdieu G 1973 *On the Four-Year Seismic Period* (Paris: Collège de France)
—— 1975 *D'Explication Optique des Points Chauds de la Terre, Consequences* (Paris: Collège de France)
Duley W W 1976 *Nature* **263** 485–6
Dyer C 1975 *Mon. Not. R. Astron. Soc.* **175** 429–47
Dyson F J 1967 *Phys. Rev. Lett.* **19** 1291–3
—— 1971 *Sci. Am.* **225** 51–9
—— 1972a *Aspects of Quantum Theory* ed A Salam and E P Wigner (London: Cambridge University Press)
—— 1972b *Inst. of Adv. Study Rep.* Princeton, NJ, USA
Eddington Sir A E 1929 *The Nature of the Physical World* (London: Cambridge University Press)
—— 1935 *New Pathways in Science* (London: Cambridge University Press)
—— 1939 *The Philosophy of Physical Science* (London: Cambridge University Press)
—— 1946 *Fundamental Theory* (London: Cambridge University Press)
Egyed L 1956 *Nature* **178** 534–5
—— 1960 *Nature* **186** 621–2
—— 1961 *Nature* **190** 1097–8
—— 1963 *Nature* **197** 1059–60

Egyed L 1969 *The Application of Modern Physics to the Earth and Planetary Interiors* ed S K Runcorn (New York: Wiley) pp65–75

Eigen M 1973 *The Physicist's Conception of Nature* ed J Mehra (Dordrecht, Holland: Reidel) pp594–633

Einstein A 1950 *The Meaning of Relativity* (Princeton, NJ: Princeton University Press)

Elder J 1967 *Nature* **214** 657–60

Ellis G F R 1971 *Proc. 47 Course of Enrico Fermi International School of Physics* (New York: Academic Press) pp104–82

Ellis H G 1974 *Found. Phys.* **4** 311–9

Elsmore B 1973 *Nature* **244** 423–4

Eötvös R, Pekar D and Fekete E 1922 *Ann. Phys.* **68** 11–66

Epstein L 1973 *Observatory* **93** 70–4

Fall S M and Jones B J T 1976 *Nature* **262** 457–60

Faulkner D J 1976 *Mon. Not. R. Astron. Soc.* **176** 621–4

Fennelly A J 1974 *Nature* **248** 221–3

Fierz M 1956 *Helv. Phys. Acta* **29** 128–34

Finzi A 1962 *Phys. Rev.* **128** 2012–5

Fish F F 1967 *Icarus* **7** 251–6

Fomalont E B and Sramek R A 1976 *Phys. Rev. Lett.* **36** 1475–8

Fowler R G and Hashemi J 1971 *Nature* **230** 518–20

Fox J G, Jacobs K C, Thomsen B and Von Hoerner S 1975 *Astrophys. J.* **201** 545–6

Fraenkel A A and Bar-Hillel Y 1958 *Foundations of Set Theory* (Amsterdam: North-Holland)

Frank F C 1971 *Geophys. J. R. Astron. Soc.* **23** 461–4

Fricke W 1972 *Rotation of the Earth* (*IAU Symp. No.* 48) ed P Melchior and S Yumi (Dordrecht, Holland: Reidel) p196

Gallant R 1963 *Nature* **200** 414–5

Gal-Or 1971 *Relativity and Gravitation* ed C G Kuiper and A Peres (London: Gordon and Breach) pp173–6

Gamow G 1948 *Nature* **162** 680–2

—— 1967a *Phys. Rev. Lett.* **19** 759–61

—— 1967b *Proc. Natn. Acad. Sci. USA* **57** 187–93

Giacaglia G E O 1972 *Rotation of the Earth* (*IAU Symp. No.* 48) ed P Melchior and S Yumi (Dordrecht, Holland: Reidel) pp165–71

Gilbert C 1956a *Mon. Not. R. Astron Soc.* **116** 684–90

—— 1956b *Mon. Not. R. Astron. Soc.* **116** 678–83

—— 1960 *Mon. Not. R. Astron. Soc.* **120** 367–86

—— 1969 *The Application of Modern Physics to the Earth and Planetary Interiors* ed S K Runcorn (New York: Wiley) pp9–18

Gilluly J 1972 *The Nature of the Solid Earth* ed E C Robertson (New York: McGraw-Hill) pp406–39

Gittus J H 1975 *Proc. R. Soc.* **A343** 155–8

Gödel K 1962 *On Formally Undecidable Propositions of Principia Mathematica and Related Systems* (London: Oliver and Boyd)

Gold T 1955 *Nature* **175** 526–9

Gold T 1966 *The Earth–Moon System* ed B G Marsden and A G W Cameron (New York: Plenum) pp93–7

Goldreich P and Toomre A 1969 *J. Geophys. Res.* **74** 2555–67

Gribbin J 1973a *Nature* **246** 453–4

—— 1973b *New Scientist* **58** 339–40

—— 1973c *New Scientist* **60** 893–5

—— 1975 *Growth Rhythms and the History of the Earth's Rotation* ed G D Rosenberg and S K Runcorn (New York: Wiley) pp413–25

Gribbin J and Plagemann S 1973 *Nature* **243** 26–7

Gross P G 1974 *Observatory* **94** 183–5

Groten E and Thyssen-Bornemisza S 1972 *Geophys. J. R. Astron. Soc.* **29** 237–9

Guinot B 1970 *Astron. Astrophys.* **8** 26–8

Gunn J E and Tinsley B M 1975 *Nature* **257** 454–7

Guthrie B N 1975 *Astrophys. Space Sci.* **43** 425–31

Hagedorn R 1974 *High Energy Astrophysics and its Relation to Elementary Particle Physics* ed K Brecher and G Setti (Cambridge, Mass.: MIT Press) pp255–96

Hales A L and Herrin E 1972 *The Nature of the Solid Earth* ed E C Robertson (New York: McGraw-Hill) pp172–215

Hallam A 1971 *Nature* **232** 180–2

Hargraves R B and Duncan R A 1973 *Nature* **245** 361–3

Hari-Dass N D H 1976 *Phys. Rev. Lett.* **36** 393–5

Harrison B K, Wakano M and Wheeler J A 1965 *Gravitation Theory and Gravitational Collapse* (Chicago: University of Chicago Press)

Harrison E R 1972 *Commun. Astrophys. Space Phys.* **4** 187–92

Hartmann W K and Larson S M 1976 *Icarus* **7** 257–60

Harwit M 1971 *Bull. Astron. Inst. Czech.* **22** 22–9

Haugan M P and Will C M 1976 *Phys. Rev. Lett.* **37** 1–4

Hawking S W 1965 *Proc. R. Soc.* **A286** 313–9

—— 1974 *Confrontation of Cosmological Theories with Observational Data* (*IAU Symp. No.* 63) ed M Longair (Dordrecht, Holland: Reidel) pp283–6

—— 1975 *Commun. Math. Phys.* **43** 199–220

Hazard C, Jauncey D L, Sargent W L W, Baldwin J A and Wampler E J 1973 *Nature* **246** 205–8

Hazel J E and Waller T R 1969 *Science* **164** 201–2

Heckmann O and Shücking E 1956 *Helv. Phys. Acta. Suppl.* **4** 114–5

Heezen B C 1960 *Sci. Am.* (Oct.) **203** 98–110

Heezen B C and Tharp M 1965 *Symp. on Continental Drift, Phil. Trans. R. Soc.* **A258** 90–106

Hellings L and Nordtvedt K 1973 *Phys. Rev.* **D7** 3593–602

Henriksen R N and Wesson P S 1978 *Astrophys. Space Sci.* **53** 429–44

Henry P S 1971 *Nature* **231** 516–8

Herrin E 1972 *The Nature of the Solid Earth* ed E C Robertson (New York: McGraw-Hill) pp216–31

Higgs P 1975 *New Scientist* **66** 60

Hill H 1975 *7th Texas Conf. on Relativistic Astrophysics, Austin, Texas*

Hill H and Stebbins R J 1975 *Astrophys. J.* **200** 471–83
Hill J M 1971 *Mon. Not. R. Astron. Soc.* **153** P7-11
Hipkin R G 1970a *Q. J. R. Astron. Soc.* **11** 43–8
—— 1970b *Palaeogeophysics* ed S K Runcorn (New York: Academic Press) pp53–9
—— 1975 *Growth Rhythms and the History of the Earth's Rotation* ed G D Rosenberg and S K Runcorn (New York: Wiley) pp315–36
Holmberg E R 1956 *Mon. Not. R. Astron. Soc.* **116** 691–8
Holmes A 1965 *Principles of Physical Geology* (London: Nelson)
Horai K I 1969 *Earth Planet. Sci. Lett.* **6** 39–42
Horák Z 1970 *Bull. Astron. Inst. Czech.* **21** 96–109
Hoyle F 1948 *Mon. Not. R. Astron. Soc.* **108** 372–82
—— 1963 *Frontiers of Astronomy* (London: Mercury)
—— 1966 *October the First is Too Late* (London: Heinemann)
—— 1969 *Q. J. R. Astron. Soc.* **10** 10–20
—— 1972a *Q. J. R. Astron. Soc.* **13** 328–45
—— 1972b *Observatory* **92** 79–82
—— 1974 *High Energy Astrophysics and its Relation to Elementary Particle Physics* ed K Brecher and G Setti (Cambridge, Mass.: MIT Press) pp297–344
—— 1975 *Astrophys. J.* **196** 661–70
Hoyle F and Narlikar J V 1964a *Proc. R. Soc.* **277** 1–23
—— 1964b *Proc. R. Soc.* **A282** 178–83
—— 1964c *Proc. R. Soc.* **A282** 184–90
—— 1964d *Proc. R. Soc.* **A282** 191–207
—— 1966a *Proc. R. Soc.* **A290** 143–61
—— 1966b *Proc. R. Soc.* **A290** 162–76
—— 1966c *Proc. R. Soc.* **A290** 177–85
—— 1966d *Proc. R. Soc.* **A294** 138–48
—— 1971 *Nature* **233** 41–4
—— 1972a *Mon. Not. R. Astron. Soc.* **155** 305–21
—— 1972b *Mon. Not. R. Astron. Soc.* **155** 323–35
—— 1972c *Cosmology, Fusion and Other Matters* ed F Reines (Bristol: Adam Hilger) pp15–28
—— 1974 *Action at a Distance in Physics and Cosmology* (San Francisco: Freeman)
Hughes D W 1974 *Nature* **253** 591–2
—— 1975a *Nature* **257** 14
—— 1975b *Nature* **258** 110–2
—— 1976 *New Scientist* **71** 64–6
Hughes D W, Robinson H G and Beltran-Lopez V 1960 *Phys. Rev. Lett.* **4** 342–4
Hundhausen A J, Bame S J and Montgomery D 1971 *J. Geophys. Res.* **76** 5145–54
Hunter J H 1969 *Mon. Not. R. Astron. Soc.* **142** 473–98
Ilič M 1974 *Preprint* Belgrade University, Yugoslavia
Infeld L 1963 *Z. Phys.* **171** 34–43

Inglis D R 1957 *Rev. Mod. Phys.* **29** 9–19

Irving E 1969 *The Application of Modern Physics to the Earth and Planetary Interiors* ed S K Runcorn (New York: Wiley) p111

Israel W and Khan K A 1964 *Nuovo Cim.* **33** 331–44

Jaakkola T 1972 *Nature* **234** 534–5

Jacobs J A 1972 *Rotation of the Earth* (*IAU Symp. No.* 48) ed P Melchior and S Yumi (Dordrecht, Holland: Reidel) pp179–81

Jacobs J A and Aldridge K D 1975 *Growth Rhythms and the History of the Earth's Rotation* ed G D Rosenberg and S K Runcorn (New York: Wiley) pp337–54

Jacobs J A and Masters G 1976 *Nature* **261** 483–4

Jeans Sir J H 1928 *Astronomy and Cosmogony* (London: Cambridge University Press)

Jeffreys Sir H 1962 *Nature* **195** 448

—— 1967 *Theory of Probability* 3rd edn (Oxford: Clarendon)

—— 1970 *The Earth* 5th edn (London: Cambridge University Press)

—— 1973 *Nature* **246** 346

—— 1975 *Q. J. R. Astron. Soc.* **16** 145–51

Johnston M J and Mauk F J 1972 *Nature* **239** 266–7

Jones B 1976 *Rev. Mod. Phys.* **48** 107–49

Jordan P 1949 *Nature* **164** 637–40

—— 1955 *Schwerkraft und Weltall* 2 Aufl. (Braunschweig: Vieweg)

—— 1959 *Z. Phys.* **157** 112–21

—— 1962a *Rev. Mod. Phys.* **34** 596–600

—— 1962b *Recent Developments in General Relativity* (New York: Pergamon) pp283–8, 289–92

—— 1964 *Proc. Lisbon Summer School on Cosmological Models, Lisboa*

—— 1967 *Z. Phys.* **201** 394–5

—— 1968 *Z. Astrophys.* **68** 201–3

—— 1969 *The Application of Modern Physics to the Earth and Planetary Interiors* ed S K Runcorn (New York: Wiley) pp55–63

—— 1971 *The Expanding Earth* (New York: Pergamon)

—— 1973 *The Physicist's Conception of Nature* ed J Mehra (Dordrecht, Holland: Reidel) pp60–70

Jordan T 1974 *J. Geophys. Res.* **79** 2141–2

—— 1975 *Nature* **257** 745–50

Kakuta C and Aoki S 1972 *Rotation of the Earth* (*IAU Symp. No.* 48) ed P Melchior and S Yumi (Dordrecht, Holland: Reidel) pp192–5

Kalitzin N St 1967 *Liège Symposia* vol 15 (Belgium: Cointe Sclessin) pp81–123

Kanasewich E R and Savage J C 1969 *The Application of Modern Physics to the Earth and Planetary Interiors* ed S K Runcorn (New York: Wiley) pp35–46

Kapp R O 1960 *Towards a Unified Cosmology* (London: Hutchinson)

Karlsson K 1971 *Astron. Astrophys.* **13** 333–5

Kaula W M 1966 *The Earth–Moon System* ed B G Marsden and A G W Cameron (New York: Plenum) pp46–51

Kaula W M 1972 *The Nature of the Solid Earth* ed E C Robertson (New York: McGraw-Hill) pp385–405

Kelly P M and Lamb H H 1976 *Nature* **262** 5

Kilmister C W and Tupper B O J 1962 *Eddington's Statistical Theory* (Oxford: Clarendon)

King J G 1960 *Phys. Rev. Lett.* **5** 562–5

King-Hele D 1970 *Nature* **226** 439–40

King-Hele D and Cook G 1973 *Nature* **246** 86–8

Kirzhnits D A 1972 *JETP Lett.* **15** 529–31

Kirzhnits D A and Linde A D 1972 *Phys. Lett.* **42B** 471–4

Klein O 1958 *Institut International Physique Solvay, 11e Congres de Physique* (Brussels: Stoops)

—— 1972 *Science* **171** 339–45

Knopoff L 1969 *Science* **163** 1277–87

Krat V A and Gerlovin I L 1974 *Astrophys. Space Sci.* **26** 521–2

—— 1975a *Astrophys. Space Sci.* **33** L5–8

—— 1975b *Astrophys. Space Sci.* **34** L11–2

Kreuzer L 1968 *Phys. Rev.* **169** 1007–12

Kundt W 1976 *Nature* **259** 30–1

Lambeck K and Cazenave A 1973 *Geophys. J. R. Astron. Soc.* **32** 79–93

Lanczos C 1957 *Rev. Mod. Phys.* **29** 337–53

Landau L 1955 *Niels Bohr and the Development of Physics* ed W Pauli (New York: McGraw-Hill) pp52–69

Landsberg P T and Bishop N T 1975a *Phys. Lett.* **53A** 109–10

—— 1975b *Mon. Not. R. Astron. Soc.* **171** 279–86

Landsberg P T and Park D 1975 *Proc. R. Soc.* **A346** 485–95

Lavrukhina A K and Ustinova G K 1971 *Nature* **232** 462–3

Lawden D F 1967 *Tensor Calculus and Relativity* 2nd edn (London: Methuen)

Lawrence J K and Szamosi G 1974 *Nature* **252** 538–9

Leader E and Williams P G 1975 *Nature* **257** 93–9

Le Denmat G, Moles M, Vigier J P and Nieto J L 1975 *Nature* **257** 773–4

Lessner G 1974 *Astrophys. Space Sci.* **30** L5–7

Lewis B M 1975 *Observatory* **95** 168–71

—— 1976 *Nature* **261** 302–4

Linde A D 1974 *JETP Lett.* **19** 183–4

Lindquist R W and Wheeler J A 1957 *Rev. Mod. Phys.* **29** 432–43

Lindsay J F and Srnka L J 1975 *Nature* **257** 776–8

Lingenfelter R E and Schubert G 1974 *Nature* **249** 820–1

Lister C B 1975 *Nature* **257** 663–5

Llewellyn-Smith C 1975 *New Scientist* **66** 74–8

Long D R 1976 *Nature* **260** 417–8

Lovell B 1971 *Q. J. R. Astron. Soc.* **12** 98–132

Lyttleton R A 1965 *Proc. R. Soc.* **A287** 471–93

—— 1970 *Adv. Astron. Astrophys.* **7** 83–145

—— 1972 *Preprint* Cambridge University

McAdoo D C and Burns J A 1975 *Earth Planet. Sci. Lett.* **25** 347–54

MacCallum M 1976 *Nature* **264** 14–5

McCrea W H 1965 *Nature* **206** 553–5
—— 1971 *Q. J. R. Astron. Soc.* **12** 140–53
—— 1975a *Observatory* **95** 13–5
—— 1975b *Nature* **255** 607–9
—— 1975c *Nature* **257** 85–6
MacDonald G J F 1964 *Advances in Earth Sciences* ed P M Hurley (Cambridge, Mass.: MIT Press) pp199–245
—— 1966 *The Earth–Moon System* ed B G Marsden and A G W Cameron (New York: Plenum) pp165–209
MacDougall J, Butler R, Kronberg P and Sandqvist A 1963 *Nature* **199** 1080
McGlynn J C, Irving E, Bell K and Pullaiah G 1975 *Nature* **255** 318–9
McKenzie D P 1966 *J. Geophys. Res.* **71** 3995–4010
—— 1972 *The Nature of the Solid Earth* ed E C Robertson (New York: McGraw-Hill) pp323–60
McNally D 1964 *Astrophys. J.* **140** 1088–99
McVittie G C 1945 *Proc. R. Soc. Edin.* **62** 147–55
—— 1956 *Astron. J.* **61** 451–62
—— 1969 *The Application of Modern Physics to the Earth and Planetary Interiors* ed S K Runcorn (New York: Wiley) pp19–28
Machado F 1967 *Nature* **214** 1317–8
—— 1975 *Geol. Rundsch* **64** 74–84
Malin S 1974 *Phys. Rev.* **D9** 3228–34
—— 1975 *Phys. Rev.* **D11** 707–10
—— 1976 *Int. J. Theor. Phys.* **14** 347–60
Malin S R C and Saunders I 1973 *Nature* **245** 25–6
Malkus W V R 1966 *The Earth–Moon System* ed B G Marsden and A G W Cameron (New York: Plenum) pp26–32
Mansfield V N 1976 *Nature* **261** 560–1
Mansfield V N and Malin S 1976 *Astrophys. J.* **209** 335–49
Maran S P and Ögelman H 1969 *Nature* **224** 349
Marsh B D 1975 *Nature* **256** 240
Martin G F and Van Flandern T C 1970 *Science* **168** 246–7
Mateo J 1972 *Rotation of the Earth* (*IAU Symp. No.* 48) ed P Melchior and S Yumi (Dordrecht, Holland: Reidel) pp185–8
Mazzullo S J 1971 *Bull. Geol. Soc. Am.* **82** 1085–6
Meadows A J 1975 *Nature* **256** 95–7
Menard H W 1965 *Symp. on Continental Drift, Phil. Trans. R. Soc.* **A258** 109–22
Meservey R 1969 *Science* **166** 609–11
Meyerhoff A A and Meyerhoff H A 1972 *J. Geol.* **80** 34–60
Meyerhoff A A and Teichert C 1971 *J. Geol.* **79** 285–321
Miller R H 1969 *Science* **164** 67–8
Milne E A 1935 *Relativity, Gravitation and World Structure* (Oxford: Clarendon)
—— 1946 *Mon. Not. R. Astron. Soc.* **106** 180–99
—— 1948 *Kinematic Relativity* (Oxford: Clarendon)
Milne E A and Whitrow G J 1938 *Z. Astrophys.* **15** 263–98

Misner C W 1968 *Astrophys. J.* **151** 431–57

Misner C W and Sharp D H 1964 *Phys. Rev.* **B136** 571–6

Morganstern R E 1971a *Nature* **232** 109–10

—— 1971b *Phys. Rev.* **D3** 2946–50

—— 1971c *Phys. Rev.* **D4** 278–82

—— 1971d *Phys. Rev.* **D4** 282–6

—— 1972 *Nature Phys. Sci.* **237** 70–1

Morrison L V 1972 *The Moon* **5** 253–64

—— 1973 *Nature* **241** 519–20

—— 1975 *Growth Rhythms and the History of the Earth's Rotation* ed
G D Rosenberg and S K Runcorn (New York: Wiley) pp445–57

Mueller I and Schwarz C R 1972 *Rotation of the Earth* (*IAUSymp. No.* 48)
ed P Melchior and S Yumi (Dordrecht, Holland: Reidel) pp68–77

Müller P M and Stephenson F R 1975 *Growth Rhythms and the History of
the Earth's Rotation* ed G D Rosenberg and S K Runcorn (New York:
Wiley) pp459–533

Munk W J 1966 *The Earth–Moon System* ed B G Marsden and A G W
Cameron (New York: Plenum) pp52–69

Munk W J and MacDonald G J F 1960 *The Rotation of the Earth* (London:
Cambridge University Press)

Murphy C T and Dicke R H 1964 *Proc. Am. Phil. Soc.* **108** 224–46

Murray B C and Malin M C 1973 *Science* **179** 997–1000

Narai H and Ueno Y 1960 *Prog. Theor. Phys. Kyoto* **24** 593–613

Narlikar J V 1974a *High Energy Astrophysics and Its Relation to Elementary
Particle Physics* ed K Brecher and G Setti (Cambridge, Mass.: MIT
Press) pp345–85

—— 1974b *J. Phys. A: Math., Nucl. Gen.* **7** 1274–82

—— 1974c *Nature* **247** 99–100

Narlikar J V and Wickramasinghe N C 1968 *Nature* **217** 1235–6

Ne'eman Y and Tauber G 1967 *Astrophys. J.* **150** 755–66

Nelson A H 1972 *Mon. Not. R. Astron. Soc.* **188** 159–75

Newton R R 1968a *Geophys. J. R. Astron. Soc.* **14** 505–39

—— 1968b *J. Geophys. Res.* **73** 3765–71

—— 1969 *Science* **166** 825–31

—— 1972a *Mem. R. Astron. Soc.* **76** 99–128

—— 1972b *Rotation of the Earth* (*IAU Symp. No.* 48) ed P Melchior and S
Yumi (Dordrecht, Holland: Reidel) pp160–1

—— 1975 *Mon. Not. R. Astron. Soc.* **169** 331–42

Ni W T 1972 *Astrophys. J.* **176** 769–96

Nicoll J F and Segal I E 1974 *Preprint* MIT, USA

Noerdlinger P D 1973 *Phys. Rev. Lett.* **30** 761–2

Nordtvedt K L 1968a *Phys. Rev.* **169** 1017–23

—— 1968b *Phys. Rev.* **170** 1186–95

—— 1969 *Phys. Rev.* **180** 1293–8

—— 1970 *Astrophys. J.* **161** 1059–67

—— 1971 *Phys. Rev.* **D3** 1683–9

—— 1972 *Science* **178** 1157–64

Nordtvedt K L 1973 *Phys. Rev.* **D7** 2347–56

Nordtvedt K L and Will C M 1972 *Astrophys. J.* **177** 775–92

O'Hora N P J 1975 *Growth Rhythms and the History of the Earth's Rotation* ed G D Rosenberg and S K Runcorn (New York: Wiley) pp427–44

O'Hora N P J and Penny C J A 1973 *Nature* **244** 426–7

Olson R H, Roberts W O and Zerefos C S 1975 *Nature* **257** 113–5

Oort J H 1970 *Astron. Astrophys.* **7** 381–404

Öpik E J 1970 *Irish Astron. J.* **9** 120–35

Oppenheimer J R and Snyder H 1939 *Phys. Rev.* **56** 455–9

Oxburgh E R 1967 *Geophys. J. R. Astron. Soc.* **14** 403–11

Pan C 1972 *Rotation of the Earth (IAU Symp. No.* 48) ed P Melchior and S Yumi (Dordrecht, Holland: Reidel) pp206–11

Pannella G 1972 *Astrophys. Space Sci.* **16** 212–37

Pannella G, MacClintock C and Thompson M N 1968 *Science* **162** 792–6

Parker L 1976 *Nature* **261** 20–3

Partridge R B 1974 *Confrontation of Cosmological Theories with Observational Data (IAU Symp. No.* 63) ed M Longair (Dordrecht, Holland: Reidel) pp157–62

Partridge R B and Wilkinson D 1967 *Phys. Rev. Lett.* **18** 557–9

Pawley G S and Abrahamsen N 1973 *Science* **179** 892–3

Pedersen G P H and Rochester M G 1972 *Rotation of the Earth (IAU Symp. No.* 48) ed P Melchior and S Yumi (Dordrecht, Holland: Reidel) pp33–8

Peebles P J E 1971 *Physical Cosmology* (Princeton, NJ: Princeton University Press)

Peebles P J E and Dicke R H 1962 *Phys. Rev.* **128** 2006–11

Penrose R 1970 *5th Texas Conf. on Relativistic Astrophysics, Austin, Texas*

Penston M 1969 *Mon. Not. R. Astron. Soc.* **144** 425–48

Peres A 1967 *Phys. Rev. Lett.* **19** 1293–4

Petrosian V 1974 *Confrontation of Cosmological Theories with Observational Data (IAU Symp. No.* 63) ed M Longair (Dordrecht, Holland: Reidel) pp31–46

Petrosian V, Bahcall J N and Salpeter E E 1969 *Astrophys. J.* **155** L57–64

Pfleiderer J 1970 *Nature* **255** 437–8

Pines D and Shaham J 1973 *Nature* **245** 77–81

Piper J 1976 *Earth Planet. Sci. Lett.* **28** 470–8

Pirani F A 1962 *Recent Developments in General Relativity* (Oxford: Pergamon) pp89–105

Pirani F A and Deser S 1965 *Proc. R. Soc.* **A288** 133–45

Pochoda P and Schwarzschild M 1964 *Astrophys. J.* **139** 587–93

Podurets M A 1964 *Sov. Astron.* **8** 19–22

Pollack H N 1969 *Science* **163** 176–7

Press F and Briggs P 1975 *Nature* **256** 270–3

Prigogine I 1973 *The Physicist's Conception of Nature* ed J Mehra (Dordrecht, Holland: Reidel) pp561–93

Prokhovnik S J 1970 *Nature* **225** 359–61

—— 1971 *Relativity and Gravitation* ed C G Kuiper and A Peres (London: Gordon and Breach) pp275–83

Proverbio E, Carta F and Mazzoleni F 1972 *Rotation of the Earth (IAU Symp. No.* 48) ed P Melchior and S Yumi (Dordrecht, Holland: Reidel) pp97–100

Proverbio E and Poma A 1975 *Growth Rhythms and the History of the Earth's Rotation* ed G D Rosenberg and S K Runcorn (New York: Wiley) pp385–95

Raine D J 1975 *Mon. Not. R. Astron. Soc.* **171** 507–28

Ramberg H 1971 *Nature* **234** 539–40

Rees M J 1972 *Commun. Astrophys. Space Phys.* **4** 179–85

Reines F and Crouch M 1974 *Phys. Rev. Lett.* **32** 493–4

Reines F and Sobel H W 1974 *Phys. Rev. Lett.* **32** 954

Rice J R and Chinnery M A 1972 *Geophys. J. R. Astron. Soc.* **29** 79–90

Riley J M 1975 *Nature* **254** 289

Roberts M S, Brown R L, Brundage W D, Rots A H, Haynes M P and Wolfe A M 1976 *Astron. J.* **81** 293–7

Robertson H P 1935 *Astrophys. J.* **82** 284–301

—— 1936a *Astrophys. J.* **83** 187–201

—— 1936b *Astrophys. J.* **83** 257–71

Robson F H 1972 *Ann. Inst. H. Poincaré* **16A** 41–50

Rochester M G 1973 *2nd GEOP Res. Conf. on the Rotation of the Earth and Polar Motion, Ohio State University, Columbus, Trans. Am. Geophys. Union* pp769–80

Rochester M G and Smylie D E 1974 *J. Geophys. Res.* **79** 4948–51

Rochester M G, Yatskiv Y S, Sasao T and Verhoogen J 1975 *Nature* **255** 655–6

Roeder R C 1967 *Astrophys. J.* **149** 131–8

Roeder R C and Demarque P R 1966 *Astrophys. J.* **144** 1016–23

Roman C 1973 *New Scientist* **57** 180–1

Roosen R G, Harrington R S, Giles J and Browning I 1976 *Nature* **261** 680–2

Rose R D, Parker H M, Lowry R A and Kulthan A R 1969 *Phys. Rev. Lett.* **23** 655–8

Rosen N 1973 *J. Gen. Relativity and Grav.* **4** 435–47

—— 1974 *Ann. Phys., NY* **84** 455–73

Rosen S P 1975 *Phys. Rev. Lett.* **34** 774–6

Rosenberg G D and Runcorn S K 1975 *Growth Rhythms and the History of the Earth's Rotation* ed G D Rosenberg and S K Runcorn (New York: Wiley) pp535–8

Rowan-Robinson M 1972 *Astrophys. J.* **178** L81–3

Roxburgh I W 1969 *The Application of Modern Physics to the Earth and Planetary Interiors* ed S K Runcorn (New York: Wiley) pp29–31

—— 1976 *Nature* **261** 301–2

Roxburgh I W and Tavakol R 1975 *Mon. Not. R. Astron. Soc.* **179** 599–610

Rubin V C, Ford W K and Rubin J S 1973 *Astrophys. J. Lett.* **183** L111–5

Rubin V C, Ford W K, Thonnard N, Roberts M S and Graham J A 1976a *Astron. J.* **81** 687–718

Rubin V C, Thonnard N, Ford W K and Roberts M S 1976b *Astron. J.* **81** 719–37

Runcorn S K 1964 *Nature* **204** 823–5

—— 1965 *Symp. on Continental Drift, Phil. Trans. R. Soc.* **A258** 228–51

—— 1966 *The Earth–Moon System* ed B G Marsden and A G W Cameron (New York: Plenum) pp82–92

—— 1968a *Continental Drift, Secular Motion of the Pole, and Rotation of the Earth* (*IAU Symp. No. 32*) ed W Markowitz and B Guinot (Dordrecht, Holland: Reidel) pp80–5

—— 1968b *Nature* **218** 459

—— 1969a *The Application of Modern Physics to the Earth and Planetary Interiors* ed S K Runcorn (New York: Wiley) pp47–51

—— 1969b *Science* **163** 1227

—— 1975 *Growth Rhythms and the History of the Earth's Rotation* ed G D Rosenberg and S K Runcorn (New York: Wiley) pp285–92

Ryall P J C and Ade-Hall J M 1975 *Nature* **257** 117–8

Sandage A 1972 *Q. J. R. Astron. Soc.* **13** 202–21

Sandage A, Tamman G A and Hardy E 1972 *Astrophys. J.* **172** 253–63

Schatten K H 1975 *Astrophys. Space Sci.* **34** 467–80

Schiff L I 1959 *Proc. Natn. Acad. Sci. USA* **45** 69–80

—— 1960 *Am. J. Phys.* **28** 340–3

Schwebel S L 1966 *Bull. Am. Phys. Soc.* **11** 95

Sciama D W 1953 *Mon. Not. R. Astron. Soc.* **113** 34–42

—— 1971 *Modern Cosmology* (London: Cambridge University Press)

—— 1972 *New Scientist* **53** 373–4

—— 1973 *The Physicist's Conception of Nature* ed J Mehra (Dordrecht, Holland: Reidel) pp17–33

Scott J 1975 *Nature* **258** 58

Scrutton C T 1965 *Palaeontology* **7** 552–8

Scrutton C T and Hipkin R G 1973 *Earth Sci. Rev.* **9** 259–74

Sedov L I 1959 *Similarity and Dimensional Methods in Mechanics* (New York: Academic Press)

Segal I E 1976 *Mathematical Cosmology and Extragalactic Astronomy* (New York: Academic Press)

Sekiguchi N 1967 *Publ. Astron. Soc. Japan* **19** 596–605

Shapiro I I, Counselman C C and King R W 1976 *Phys. Rev. Lett.* **36** 555–8

Shapiro I I, Smith W B, Ash M B, Ingalls R P and Pettengill G H 1971 *Phys. Rev. Lett.* **26** 27–30

Shatzman E 1966 *The Earth–Moon System* ed B G Marsden and A G W Cameron (New York: Plenum) pp12–25

Sherwin C W, Frauenhelder H, Garwin E L, Lüsher E, Margulies S and Peacock R N 1960 *Phys. Rev. Lett.* **4** 399–401

Shlyakhter A I 1976 *Nature* **264** 340

Skilling J 1975 *Nature* **258** 687–8

Slater N B 1957 *Development and Meaning of Eddington's Fundamental Theory* (London: Cambridge University Press)

Smith P J 1975a *Nature* **254** 386

—— 1975b *Nature* **256** 538–40

—— 1976 *Nature* **260** 97

Smylie D E and Mansinha L 1971 *Nature* **232** 621–2
Solheim J E, Barnes T G and Smith H J 1976 *Astrophys. J.* **209** 330–4
Spall H 1972 *Nature* **236** 119–21
Stacey F D 1963 *Nature* **197** 582–3
—— 1975 *Nature* **255** 44–5
Steenbeck M and Helmis G 1975 *Geophys. J. R. Astron. Soc.* **41** 237–44
Stehli F G 1973 *Earth Sci. Rev.* **9** 1–18
Steigman G 1976 *Nature* **261** 479–80
Steiner J 1967 *J. Geol. Soc. Aust.* **14** 99–131
Steiner M B 1975 *Nature* **254** 107–9
Stephenson F R 1972 *PhD Thesis* Department of Geophysics, University of Newcastle-upon-Tyne, UK
—— 1976 *Nature* **259** 101–2
Stewart A D 1970 *Palaeogeophysics* ed S K Runcorn (New York: Academic Press) pp415–34
—— 1972 *Nature* **235** 322–4
Stewart I C 1976 *Geophys. J. R. Astron. Soc.* **46** 505–11
Stothers R 1976 *Nature* **262** 477–9
Swann W F G 1927 *Phil. Mag.* **3** 1088–136
Synge J L 1966 *Relativity: The General Theory* (Amsterdam: North-Holland)
Takeuchi H and Sugi N 1972 *Rotation of the Earth* (*IAU Symp. No.* 48) ed P Melchior and S Yumi (Dordrecht, Holland: Reidel) pp212–4
Tamman G A 1974 *Confrontation of Cosmological Theories with Observational Data* (*IAU Symp. No.* 63) ed M Longair (Dordrecht, Holland: Reidel) pp47–59
Tanguy J C and Wilson R L 1973 *Phil. Trans. R. Soc.* **A274** 163
Tarling D H 1975 *Growth Rhythms and the History of the Earth's Rotation* ed G D Rosenberg and S K Runcorn (New York: Wiley) pp397–412
Teller E 1948 *Phys. Rev.* **73** 801–9
—— 1972 *Cosmology, Fusion and Other Matters* ed F Reines (Bristol: Adam Hilger) pp60–6
Termier H and Termier G 1969 *The Application of Modern Physics to the Earth and Planetary Interiors* ed S K Runcorn (New York: Wiley) pp89–101
Thomas C and Briden J C 1976 *Nature* **259** 380–2
Thompson I H and Whitrow G J 1967 *Mon. Not. R. Astron. Soc.* **136** 207–17
—— 1968 *Mon. Not. R. Astron. Soc.* **139** 499–513
Thorne K S and Dykla J J 1971 *Astrophys. J. Lett.* **166** L35–8
Thorne K S and Will C M 1971 *Astrophys. J.* **163** 595–610
Tinsley B M 1972 *Astrophys. J. Lett.* **174** L119–21
Toomre A 1966 *The Earth–Moon System* ed B G Marsden and A G W Cameron (New York: Plenum) pp33–45
Towe K M 1975 *Nature* **257** 115–6
—— 1976 *Nature* **261** 438
Trautman A 1964 *Brandeis Summer Sch. Theor. Phys., Lectures on General Relativity* vol 1 ed A Trautman *et al* (Englewood Cliffs, NJ: Prentice-Hall) pp1–248

Trümper M 1963 *Z. Phys.* **173** 422–7

Turcotte D L, Nordman J C and Cisne I L 1974 *Nature* **251** 124–5

Uhlenbeck G E 1973 *The Physicist's Conception of Nature* ed J Mehra (Dordrecht, Holland: Reidel) pp501–13

Ulam S M 1972 *Cosmology, Fusion and Other Matters* ed F Reines (Bristol: Adam Hilger) pp272–9

Ulrych T 1972 *Nature* **235** 218–9

Urey H C 1952 *The Planets* (London: Oxford University Press)

Van Andel S I and Hospers J 1969 *The Application of Modern Physics to the Earth and Planetary Interiors* ed S K Runcorn (New York: Wiley) pp113–21

Van Flandern T C 1975 *Mon. Not. R. Astron. Soc.* **170** 333–42

Van Hilten D 1965 *Nature* **200** 1277–9

Vartanyan Y L 1966 *Astrofizika* **2** 45–52

Veizer J 1971 *Nature* **229** 480–1

Veltman M 1975 *Phys. Rev. Lett.* **34** 777

Verhoogen J 1974 *Nature* **249** 334–5

Verosub K L 1975 *Nature* **253** 707–8

Vinti J P 1974 *Mon. Not. R. Astron. Soc.* **169** 417–27

Vitello P and Salvati M 1976 *Phys. Fluids* **19** 1523–31

Von Hoerner S 1973 *Astrophys. J.* **181** 261–5

Von Weizächer C F 1973 *The Physicist's Conception of Nature* ed J Mehra (Dordrecht, Holland: Reidel) pp635–67

Waldbaum D R 1971 *Nature* **232** 545–7

Walker A G 1935 *Mon. Not. R. Astron. Soc.* **95** 263–9

—— 1936 *Proc. Lond. Math. Soc.* **42** 90–127

—— 1945 *Proc. R. Soc. Edin.* **45** 164–74

Ward M A 1963 *Geophys. J. R. Astron. Soc.* **8** 217–30

Wataghin G 1972 *Cosmology, Fusion and Other Matters* ed F Reines (Bristol: Adam Hilger) pp48–55

Wdowczyk J and Wolfendale A W 1975 *Nature* **258** 217–8

Weertman J 1976 *Nature* **261** 17–20

Weinberg S 1972 *Gravitation and Cosmology* (New York: Wiley)

—— 1974 *Phys. Rev.* **D9** 3357–78

Weinstein D H and Keeney J 1973a *Nature* **244** 83–4

—— 1973b *Lett. Nuovo Cim.* **8** 299–307

—— 1974 *Nature* **247** 140

—— 1975 *Growth Rhythms and the History of the Earth's Rotation* ed G D Rosenberg and S K Runcorn (New York: Wiley) pp377–84

Wells J W 1963 *Nature* **197** 948–50

Wesson P S 1971 *Nature* **232** 251–2

—— 1972a *J. Geol.* **80** 185–97

—— 1972b *Bull. Am. Ass. Petrol. Geol.* **56** 2127–49

—— 1973 *Q. J. R. Astron. Soc.* **14** 9–64

—— 1974a *Space Sci. Rev.* **15** 469–82

—— 1974b *Plate Tectonics—Assessments and Reassessments* ed C F Kahle (Tulsa: American Association of Petroleum Geologists) pp146–54

Wesson P S 1974c *Plate Tectonics—Assessments and Reassessments* ed C F
Kahle (Tulsa: American Association of Petroleum Geologists) pp448–62
—— 1975a *Astrophys. Space Sci.* **36** 363–82
—— 1975b *Growth Rhythms and the History of the Earth's Rotation* ed G D
Rosenberg and S K Runcorn (New York: Wiley) pp353–74
Wesson P S and Lermann A 1976 *Astron. Astrophys.* **53** 383–8
Weyl H 1922 *Space–Time–Matter* (London: Methuen)
Weymann R J, Bovoson T and Scargle J D 1978 *Astrophys. Space Sci.* **53**
265–78
Wheeler J A 1962 *Geometrodynamics* (New York: Academic Press)
—— 1964a *Relativity, Groups and Topology* ed C and B De Witt (London:
Gordon and Breach) pp317–22
—— 1964b *Relativistic Theories of Gravitation* ed L Infeld (Oxford: Perga-
mon) pp223–68
Whitrow G J 1961 *The Natural Philosophy of Time* (London: Nelson)
Wilkinson D H 1958 *Phil. Mag.* **3** 582–5
—— 1975 *Nature* **257** 189–93
Will C M 1971a *Astrophys. J.* **163** 611–28
—— 1971b *Astrophys. J.* **165** 409–12
—— 1971c *Astrophys. J.* **169** 125–40
—— 1971d *Astrophys. J.* **169** 141–55
—— 1972 *Physics Today* (Oct.) **25** 23–9
—— 1976 *Astrophys. J.* **204** 224–34
Will C M and Nordtvedt K 1972 *Astrophys. J.* **177** 757–74
Williams E R, Faller J E and Hill H 1971 *Phys. Rev. Lett.* **26** 721–4
Williams G E 1974 *J. Geol. Soc.* **130** 599–601
—— 1975 *Earth Planet. Sci. Lett.* **26** 361–9
Williams I P and Cremin A W 1968 *Q. J. R. Astron. Soc.* **9** 40–62
Wilson J T 1960 *Nature* **185** 880–2
—— 1961 *Nature* **192** 125–8
Wilson O C 1966 *Astrophys. J.* **144** 695–708
Wilson R L 1970 *Geophys. J. R. Astron. Soc.* **19** 417–37
Wolfe A M 1970 *Astrophys. J. Lett.* **159** L61–7
Wolfe A M, Brown K L and Roberts M S 1976 *Phys. Rev. Lett.* **37** 179–81
Wolfe A M and Burbidge G R 1969 *Astrophys. J.* **156** 345–71
Wood K D 1972 *Nature* **240** 91–3
Woollard G P 1972 *The Nature of the Solid Earth* ed E C Robertson (New
York: McGraw-Hill) pp463–505
Yukutake T 1972 *Rotation of the Earth* (*IAU Symp. No.* 48) ed P Melchior and
S Yumi (Dordrecht, Holland: Reidel) pp229–30
Zeldovitch Y B 1974 *Confrontation of Cosmological Theories with Observa-
tional Data* (*IAU Symp. No.* 63) ed M Longair (Dordrecht, Holland:
Reidel) pp329–33
Zeldovitch Y B and Novikov I D 1971 *Relativistic Astrophysics* vol 1
(Chicago: University of Chicago Press)

Index

240